ALGEBRA

Part III. Functions plus

Statistics & Probability

Lessons for Self-Study with Test Preparation

Build Your Self-Confidence and Enjoyment of Math!

All about Functions in Algebra and

Statistics & Probability

with a Comprehensive Solutions Manual

Aejeong Kang

MathRadar

Send all inquiries to:

MathRadar, LLC
5705 Spring Hill Dr.
Mckinney, Texas 75072

Visit www.mathradar.com for more information and a sneak preview of the MathRadar series of math books.

Send inquires via email at info@mathradar.com

Algebra, Part III: Function and Statistics & Probability

ISBN-13: 978-0-9893689-4-0

ISBN-10: 0989368947

Printed in the United States of America.

Preface

I wrote these books because I am a mother and I have a strong academic background in mathematics. I have a BS degree in Mathematics and Master's degree in Mathematics as well. I have completed Ph.D. program in Biostatistics.

After receiving the big blessing of our first child, a daughter, I decided to forgo my personal career goals to become a full-time mother. When our daughter entered 7th grade, that meant lots of help with her study of math-my passion. However, I struggled to find good math books that would help her understand difficult concepts both clearly and quickly. After the conversation with my husband and (now two) children, I decided that the best way to help my children was by writing math books for them myself. They wholeheartedly agreed.

That's why I've been able to pour all my knowledge, energy, and soul into these books. Because I'm a mom, I would do anything for my children. Thanks to my family's endless support, I wrote them four books, designed for use in junior high and high-school (partially) mathematics.

And that would have been the end of my journey, but my husband and children insisted that I share my work outside of our family. They encouraged me to make my work available to other parents looking, as I was, for well-written, great mathematics books for their children.

So I finally decided to publish these books. I do so with the hope that they will help your children find success and confidence in learning and studying mathematics.

But I would never have begun or finished this project without the support of my family. Kyungwan, Nichole, and Richard, you are my world. Thank you.

Introduction

✔ *After reading several pages of explanation/description about a certain mathematical concept, you still don't get it.*

✔ *You have worked on many related problems to understand mathematical concepts, but you still feel completely lost in the mathematical jungle.*

✔ *You bought a math book with good reviews, but it only offers short answers without detailed solutions. You feel confused and frustrated.*

✔ *You've tried multiple learning math books, but you've still not getting good grades in math. It seems like math is just not for you.*

If any one of these situation sound familiar, the MathRadar series will help you escape!

Everyone has different learning abilities and academic skill. The MathRadar series is written and organized with emphasis on helping each individual study mathematics at his/her own pace.

In the case of Algebra, each book covers all the topics required in each field of Algebra. These fields were systematically subdivided into Part I, Part II, and Part III.

Algebra, Part I covers information about **Number Systems** from natural numbers to real numbers.

Level I	for **grades 6~8**	: Chapter 1 and Chapter 2
Level II	for **grades 7~9**	: Chapter 3
Level III	for **grades 8~10**	: Chapter 4

Algebra, Part II covers information about **Expressions** dealing with equations and inequalities.

Level I	for **grades 6~8**	: Chapter 1 and Chapter 2
Level II	for **grades 7~9**	: Chapter 3, Chapter 4, and Chapter 5
Level III	for **grades 8~10**	: Chapter 6, Chapter 7, and Chapter 8

Algebra, Part III covers information about __Functions__, while also including __Statistics and Probability__.

Level I	for **grades 6~8**	: Chapter 1
Level II	for **grades 7~9**	: Chapter 2
Level III	for **grades 8~10**	: Chapter 3

Statistics : Chapter 1, Chapter 2, and Chapter 3
Probability : Chapter 4

Each book consists of clean and concise summaries, callouts, additional supporting explanations, quick reminders and/or shortcuts to facilitate better understanding.

With the numerous examples and exercises, students can check their comprehension levels with both basic and more advanced problems.

Each book includes **Solutions Manual.** The solutions manual makes it possible for students to study difficult concepts on their own. With the solutions manual, students will be able to better understand how to solve problems through step-by-step for each problem.

Geometry has also been systematically subdivided so that students can easily grasp geometry concepts. Each concept is thoroughly explained with step-by-step instruction and detailed proofs.

Carry the MathRadar series with you!

Work on them anytime and anywhere!

Finally, you can start to enjoy mathematics!

Whether you are struggling or advanced in your math skills, the MathRadar series books will build your self-confidence and enjoyment of math.

I hope Math Radar is what you need and will be a great tool for your hard work.

Your comments or suggestions are greatly appreciated.

Please visit my website at www. mathradar.com or email me at aejeong@mathradar.com

Thank you very much. And remember, math can be fun!

Aejeong Kang

How to use the MathRadar series

Once you finish level I of the MathRadar Algebra book of your choice, you will have 2 options to choose from.

Option 1. Keep working on level II and III of the same book
to complete your study up to the beginning of high school level.

Option 2. Start working on level I in other MathRadar books
to complete your current level in other areas of Algebra.

Since the MathRadar Series is written to meet with everyone's different learning abilities and academic skills, and not for the grade level or age, you can build your progress quickly with strong confidence. Once you complete all the levels of the MathRadar series, you can see yourself on highly advanced level of mathematics. Then you can continue to learn anything very easily with a strong foundation and base.
The MathRadar series of math books will be the beneficial and valuable resources towards your goal.

Level I :   Grades 6-8

Level II :   Grades 7-9

Level III :   Grades 8-10

Algebra - Part III : Functions
Level I   : Chapter 1
Level II   : Chapter 2
Level III : Chapter 3

➡ FUNCTIONS

TABLE OF CONTENTS

Chapter 1 Functions level I

Chapter 2   Linear Functions level II

Chapter 3   Quadratic Functions

level III

➡ **STATISTICS AND PROBABILITY**

TABLE OF CONTENTS

Chapter 1   Basic Statistical Graphs

Chapter 2   Descriptive Statistics

Chapter 3   The Concept of Sets

Chapter 4　Probability

Solutions Manual for Functions in Algebra

Solutions Manual for Statistics and Probability

Index

Algebra

**Part III
Functions**

Chapter 1

**Functions
Level I**

CHAPTER 1

Chapter 1 Functions

Chapter1. Functions

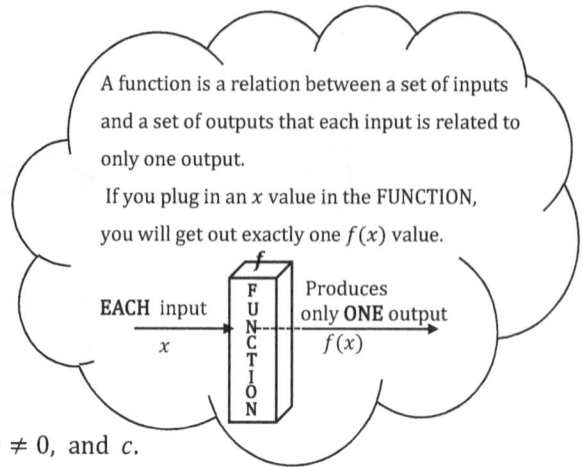

1-1 Functions

1. Function

A *standard form* of a linear equation is formed by

$$ax + by = c \text{ for any constants } a \neq 0, b \neq 0, \text{ and } c.$$

Solving the equation for y, we get the y in terms of x.

For the relationship between x and y, $y = ax + b, a \neq 0$, y is called a *function* of x and is written

$$y = f(x)$$

Each value of y is dependent on the chosen value of x, that is each value of x is assigned exactly one value of y. So, we call x the *independent variable* and y the *dependent variable*.

Note : A function is a relationship in which each element of the domain is paired with exactly one element of the range.

y *is a function of* x . $\underset{\text{represented by}}{\Longleftrightarrow}$ $f : x \longrightarrow y$ or $x \xrightarrow{f} y$ or $y = f(x)$

2. Domain and Range

The *domain* of a function is the set of all input values for which $y = f(x)$ is defined.

The *range* of a function is the set of all output values for $y = f(x)$.

Example

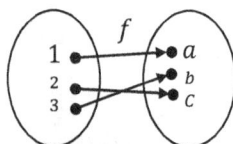

Domain : {1,2,3}

Range of $f(x)$: $\{a, b, c\}$

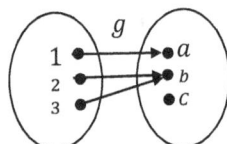

Domain : {1,2,3}

Range of $g(x)$: $\{a, b\}$

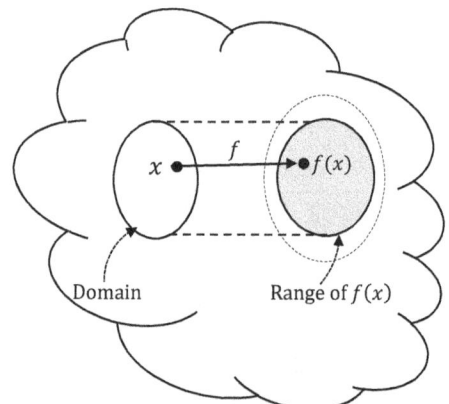

Note

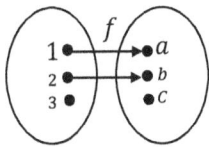

Not a function
($\because f(3)$ does not exist.)

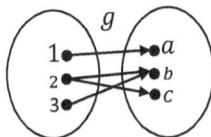

Not a function
($\because g(2)$ has 2 values.)

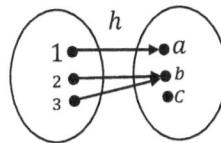

A function
($\because$ Each value of x is assigned exactly one $h(x)$.)

1-2   Graphing and Solving Functions

y is a function of x if any value of y is determined by the value of x.

1.   Coordinate Planes

(1) Coordinates

A *coordinate* is the number of a point assigned on a number line.

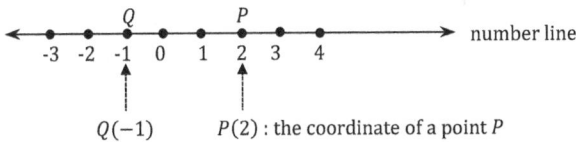

$Q(-1)$ $P(2)$: the coordinate of a point P

Note : A number line is one-dimensional, and a coordinate plane is two-dimensional. So, an equation with 1 variable is graphed on a number line and an equation with 2 different variables is graphed on a coordinate plane.

(2) Coordinate Planes

When two number lines are perpendicular to each other at the zero points of the two lines,

1)   their intersection is called the *origin*,

2)   the horizontal number line is called the *x-axis*,

3)   the vertical number line is called the *y-axis*,

4)   the plane formed by the *x*-axis and the *y*-axis is called the *coordinate plane*.

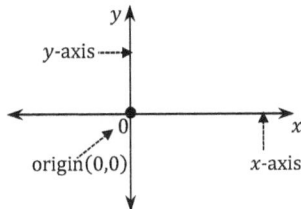

The coordinate of origin is (0,0).

Note.   Each point P in the coordinate plane is assigned a pair of numbers.

(3) 4 Quadrants

4 Quadrants are four rectangular regions in a coordinate plane which are labeled with Roman numerals I, II, III, and IV. The coordinate plane is divided by the two axes into 4 quadrants, which are named in counterclockwise order.

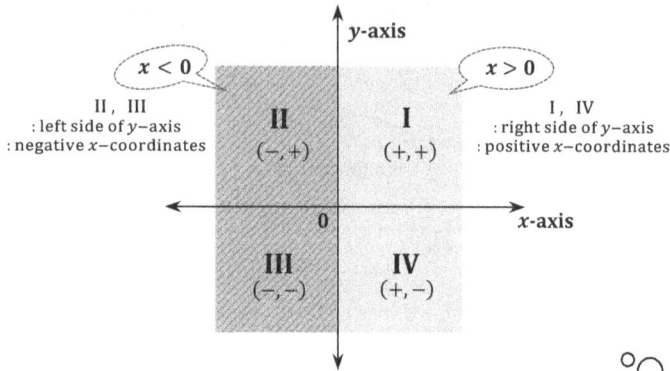

The x-axis and y-axis do not belong to any quadrants.

I, II : above the x-axis
: positive y-coordinates ; $y > 0$
III, IV : below the x-axis
: negative y-coordinates ; $y < 0$

(4) Ordered Pairs

An ordered pair (a, b) is the coordinates of the point P.

1) a is the number of horizontal units moved from 0.

2) b is the number of vertical units moved from 0.

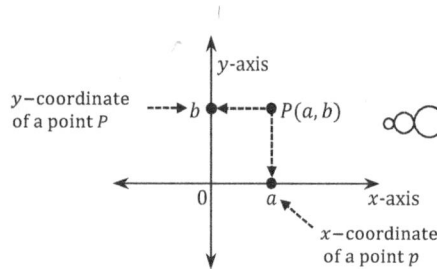

$(x, 0)$; The coordinate of a point on the x-axis
$(0, y)$; The coordinate of a point on the y-axis

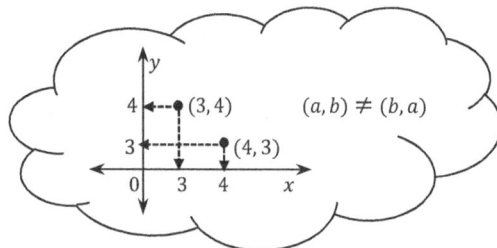

(5) Symmetric Transformation

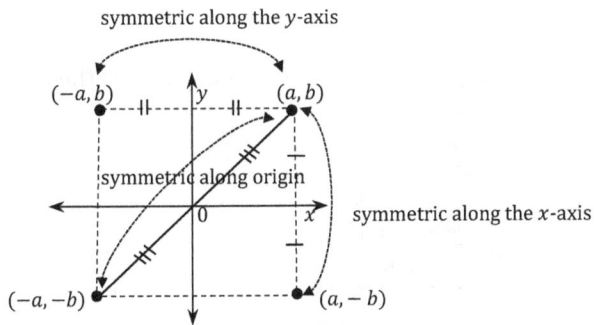

1) Symmetry of a point (a, b) along the x-axis ; $(a, -b)$: opposite sign of y

2) Symmetry of a point (a, b) along the y-axis ; $(-a, \ b)$: opposite sign of x

3) Symmetry of a point (a, b) along the origin ; $(-a, -b)$: opposite signs of x and y

Note : (1) *The middle point between the two points* $A(a_1, b_1)$ *and* $B(a_2, b_2)$ *is*

$$\left(\frac{a_1 + a_2}{2}, \frac{b_1 + b_2}{2} \right)$$

(2) *The distance between the two points* $A(a_1, b_1)$ *and* $B(a_2, b_2)$ *is*

$$\sqrt{(a_2 - a_1)^2 + (b_2 - b_1)^2} \quad (\ by \ Pythagorean \ Theorem)$$

2. Graphing Functions

(1) Graphing $y = ax, (a \neq 0)$

The graph of this function is known as a direct variation : a straight line through the origin (0,0).

1) $a > 0$

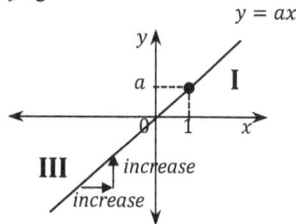

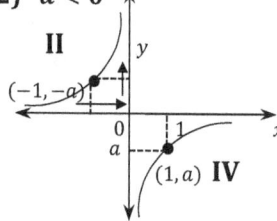

As the x values increase (rise)
from left to right,
the y values increase as well
in the coordinate plane.

2) $a < 0$

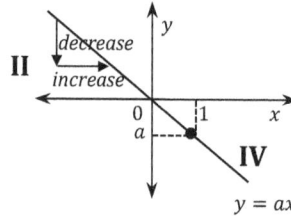

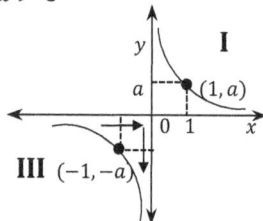

As the x values increase (rise)
from left to right,
the y values decrease(decline)
in the coordinate plane.

$$y = ax, (a \neq 0) \Rightarrow \frac{y}{x} = a \text{ (The ratio is a constant.)}$$

$y = 2x$ close to the y-axis

$y = \frac{1}{2}x$: close to x-axis $\quad |2| > \left|\frac{1}{2}\right|$

$y = -2x$ close to the y-axis

$y = -\frac{1}{2}x$: close to x-axis $\quad |-2| > \left|-\frac{1}{2}\right|$

To graph this type of function, identify ordered pairs on the coordinate plane and connect them with a straight line.

(2) Graphing $y = \frac{a}{x}, (a \neq 0)$

The graph of this function is known as an inverse variation : a pair of smooth curves which are symmetric along the origin (0,0).

1) $a > 0$

As the x values increase (rise)
from left to right,
the y values decrease
in the coordinate plane.

2) $a < 0$

As the x values increase (rise)
from left to right,
the y values increase as well
in the coordinate plane.

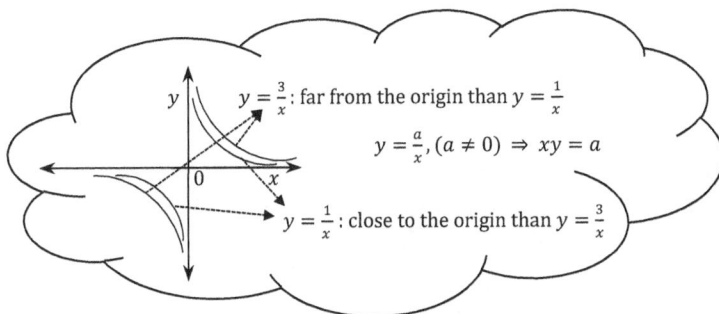

Note : From the graph of $y = \dfrac{a}{x}, (a \neq 0)$,

(1) As the value of x moves away from 0, the graph moves closer to the x-axis, but it never crosses the x-axis.

(2) As the value of x approaches 0, it gets closer to the y-axis, but it never crosses the y-axis.

3. Steps for Solving Word Problems

(1) Determine the two variables x and y.

(2) Express the relationship between x and y as a function .

For example, $y = ax, (a \neq 0)$ or $y = \dfrac{a}{x}, (a \neq 0)$

(3) Find the solution for the equation.

(4) Check the solution.

Solution Problems: The salt refers to as the solute, the water is the solvent, and the resulting mixture as the solution (water+salt). The amount (the concentration) of salt in the solution expresses how salty the salt water is. $$\text{Concentration} = \frac{\text{The amount of salt (solute)}}{\text{The total amount of solution}}$$ Concentration is normally expressed as a percent (%), multiplied by 100 . **The amount of salt (solute)** **= (Concentration) × (The total amount of solution)**	Distance, Rate, and Time Problems: **Distance = Rate × Time** $\text{Rate} = \dfrac{\text{Distance}}{\text{Time}}$, $\text{Time} = \dfrac{\text{Distance}}{\text{Rate}}$ If the rate is in miles per hour, then the distance must be in miles and the time in hours. If the time is in minutes, convert it to hours (dividing by 60) to find the distance in miles. *Match the units!!*

Exercises

#1. The domain of a function $f(x) = -2x + 3$ is $\{0, 1, 2, 3\}$. Find the range of $f(x)$.

#2. The range of a function $f(x) = 2x$ is $\{-8, 0, 4, 8\}$. Find the domain of $f(x)$.

#3. Find the domain and range of the equation $y = |x| - 3$.

#4. The range of a function $g(x) = ax$ is $\{-2, 0, 2\}$ when $g(2) = -1$. Find the domain of the function.

#5. $A = \{(-3,1),(-2,2),(-2,3),(-1,4),(0,5),(1,5)\}$ is the set of ordered pairs. Is this relationship a function?

#6. For a function $f(x) = ax$, $f(3) = -4$. Find the value of $f(9)$.

#7. Find the value of $f(3) - f(2) + f(4)$ for the function $f(x) = \frac{3}{x}$.

#8. For the two functions $f(x) = ax + 2$ and $g(x) = \dfrac{b}{x} - 2$, $f(1) = g(-1) = 3$. Find the value of $a + b$.

#9. For the two functions $f(x) = \dfrac{a}{x} + 2$ and $g(x) = -\dfrac{3}{x} + 5$, $3f(-2) = 2g(-3)$.

Find the value of b which satisfies $f(b) = g(b)$.

#10. Identify functions.

(1)

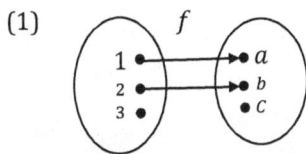

(2)

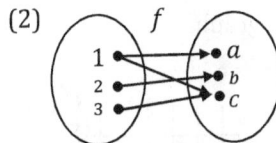

(3)

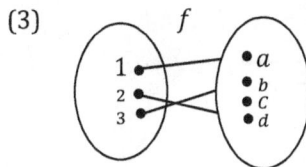

(4)

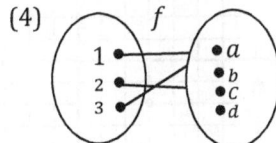

(5)

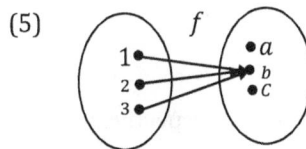

(6)

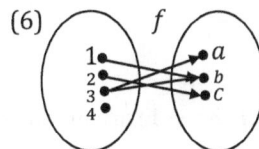

#11. For the function $f(3x - 2) = 2x - a$, $f(4) = 3$. Find the value of $f(1)$.

#12. For the two functions $f(x) = 2ax$ and $g(x) = \frac{2}{x} - 1$, $g(f(2)) = 3$. Find the value of a.

#13. Plot the following ordered pairs on the graph.

(1) $A(2, 3)$

(2) $B(-2, 3)$

(3) $C(2, -3)$

(4) $D(-5, 5)$

(5) $E(0, 5)$

(6) $F(4, 0)$

(7) $G(-3, 0)$

(8) $H(0, -7)$

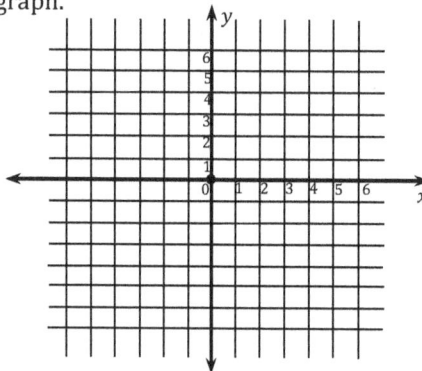

#14. Find the coordinates for each point on the graph.

(1) A

(2) B

(3) C

(4) D

(5) E

(6) F

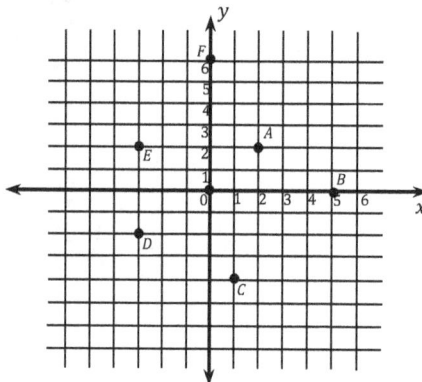

#15. Two points $P(a + 2, 4 - 2a)$ and $Q(2 - 2b, 3b + 1)$ are on the x-axis and y-axis respectively.

Find the value of $a + b$.

#16. Find the length of the segment between the two points.

(1) $A(1, 2)$ and $B(1, -2)$

(2) $C(0, 3)$ and $D(-3, 3)$

(3) $P(-3, 0)$ and $Q(5, 0)$

(4) $S(-5, 0)$ and $T(-5, -6)$

#17. A point (a, b) is in the second quadrant of the coordinate plane.

Name the quadrant containing the following points.

(1) $(a, -b)$

(2) $(-b, a)$

(3) (b, a)

(4) $(-a, -b)$

(5) $(-a, b)$

(6) $(-b, -a)$

(7) (ab, a^2)

(8) $(-a, -ab)$

#18. Point B is reflected through the origin to point $A(3, 4)$. Point C is obtained by reflecting point B across the y-axis. Find the area of a triangle $\triangle ABC$.

#19. Point $C(4, b)$ is the midpoint of Points $A(-2, 3)$ and $B(a, 9)$. Find the value of $a - b$.

#20. Which graphs are functions?

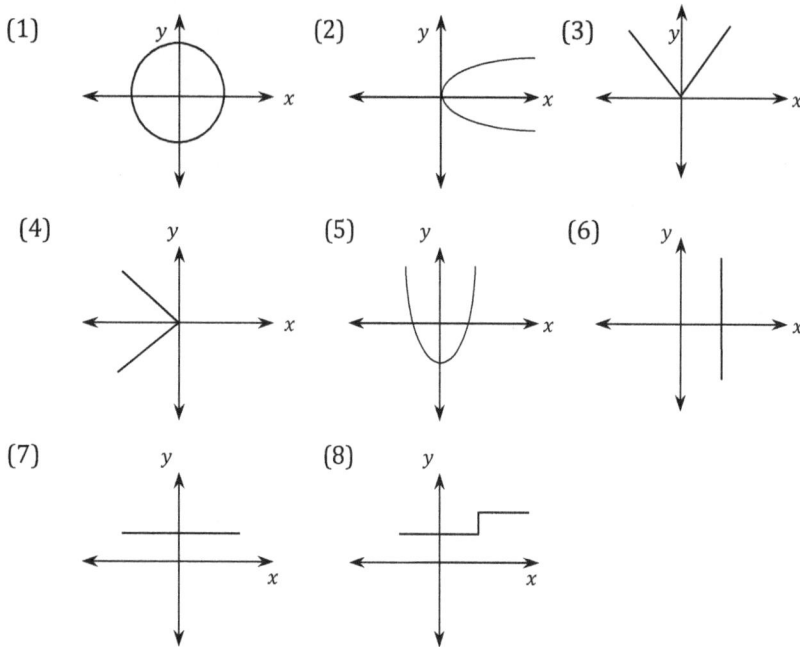

(1)

(2)

(3)

(4)

(5)

(6)

(7)

(8)

#21. Identify the function of the form $y = ax$ which passes through the origin and $(3, -4)$.

#22. Identify the functions of the form $y = ax$ or $y = \frac{a}{x}$ for the following graphs.

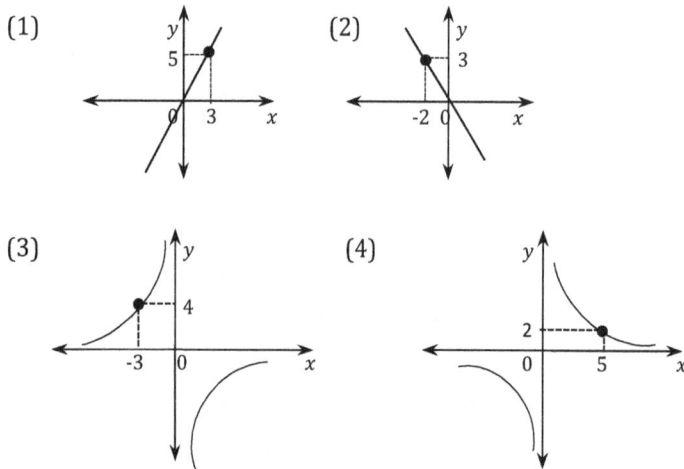

(1)

(2)

(3)

(4)

#23. Find the functions for the data in the tables below.

(1)

x	−4	−2	−1	1	2	4
y	−1	−2	−4	4	2	1

(2)

x	−2	−1	0	1	2	3
y	−1	1	3	5	7	9

(3)

x	1	2	3	4	6	12
y	12	6	4	3	2	1

(4)

x	−4	−2	0	2	4	6
y	2	1	0	−1	−2	−3

#24. The function $f(x) = -\dfrac{3}{2}x$ passes through a point $(a + 1, 2a - 3)$. Find the value of a.

#25. The function $y = ax$ passes through a point $(3, -15)$ and $(b, 10)$. Find the value of $a - b$.

#26. For any constants a and b, the function $f(x) = \dfrac{2a}{x}$ passes through the points $(-2, 8)$ and $(4, b)$.

Find the value of $a + b$.

#27. For any constants $a, b,$ and $c,$ the function $f(x) = \frac{a}{x}$ passes through the points $(b, 1), (1, c),$ and $(3, -1)$.

Find the value of $a + b + c$.

#28. Two functions $f(x) = ax$ and $g(x) = \frac{b}{x}$ meet at the points $(3, 9)$ and $(-3, c)$. Find the value of $a + b + c$.

#29. Two functions $y = -ax$ and $y = -\frac{2}{x}$ meet at Point $A(b, 8)$. Find the value of ab.

#30. Find the function of the form $y = ax$ or $y = \frac{a}{x}$ for the graph.

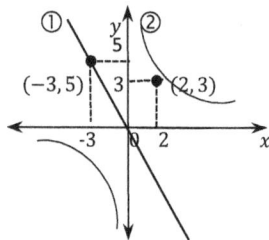

#31. The function $y = 3x$ passes through the two points, origin and A. The area of the triangle $\triangle OAB$ is 54.

Find the coordinate of Point A.

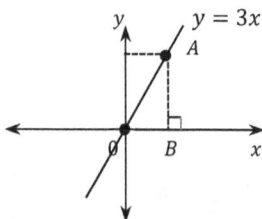

#32. Two points $P(3, a)$ and $Q(3, b)$ are on the graph $y = 3x$ and $y = -x$, respectively.

Find the area of the triangle $\triangle OPQ$.

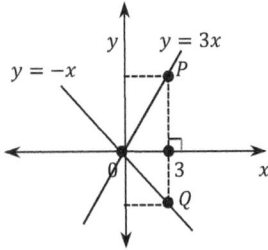

#33. Richard rides his bike from home to a park 5 miles away at a speed of x miles per hour for y hours. Find the relationship between x and y.

#34. Which one is not a function?

(1) The sum of two variables x and y is 5.

(2) The variable y is half of the variable x.

(3) The perimeter (y inch) of a rectangle with one side of length (x inch).

(4) 10 miles at a speed of x miles per hour for y minutes.

#35. A building needs to be painted. It takes 30 hours for 5 workers to finish the job. If the job has to be finished in 6 hours, how many workers are needed?

#36. Nichole wants to make a vegetable garden with an area of 200 square feet. Find the relation between the length (x feet) and width (y feet).

#37. The distance from A to B is 10 miles. Nichole drives at a speed of 35 miles per hour from A to B, and Richard drives at a speed of 25 miles per hour from B to A at the same time. How long will it take before they meet each other?

#38. Richard drives to a post office at a speed of 50 miles per hour. 5 minutes later, Nichole drives to the post office at a speed of 60 miles per hour. How long will it take before Nichole meets Richard?

#39. Richard drives to school at a speed of 40 miles per hour and returns back home at a speed of 30 miles per hour. Coming home, it takes him 10 more minutes than going school. How far is it from Richard's home to school?

#40. Nichole rides her bike halfway to school at 20 mph. She drives her car the rest of the way at 40 mph. Find Nichole's average speed to school.

#41. x ounces of a y% salt solution contains 3 ounces of salt . Find the relationship between x and y.

#42. Nichole wants to buy some books at a bookstore which are all the same price. If she buys 3 books, then she will be $2.50 short. If she buys 2 books, then she will have $5.00 left over. How much money does she have?

#43. 3 machines can do 5 jobs in 4 hours. How many hours will it take for 4 machines to do 6 jobs?

Algebra

Part III
Functions

Chapter 2

Linear Functions
Level II

CHAPTER 2

Chapter 2 Linear Functions

2-1 Linear Functions and their Graphs

1. **Linear Functions**

2. **Graphing $y = ax + b,\ a \neq 0$**

    (1) Graphing $y = ax + b,\ a \neq 0$ using Translation

    (2) Intercepts

    (3) Graphing $y = ax + b,\ a \neq 0$ using Intercepts

3. **Horizontal and Vertical Lines**

4. **The Slope of a Line**

    (1) The Concept of Slope

    (2) Properties of Slope

    (3) Graphing $y = ax + b,\ a \neq 0$ using Slope and y-intercept

2-2 Lines and their Equations

1. **Methods of Creating the Equations of Lines**

2. **Equations of Lines**

3. **Parallel and Perpendicular Lines**

4. **Translation and Symmetry for $y = ax,\ a \neq 0$**

2-3 Solving Equations by Graphing

2-4 Graphing with Absolute Values

Chapter 2. Linear Functions

2-1 Linear Functions and their Graphs

1. Linear Functions

> A linear function is a function that can be graphically represented in a coordinate plane by a straight line.

A function f is called a *linear function of x* if it is formed by

$$f(x) = ax + b \quad \text{(function form) or}$$
$$y = ax + b \quad \text{(equation form)},$$
$$\text{where } a(\neq 0), b \text{ are constants}$$

> Linear :
> The highest power in a polynomial is 1.

Note : $f(x) = ax + b$, $a \neq 0$ is a first-degree polynomial function or linear function.

$f(x) = ax^2 + bx + c$, $a \neq 0$ is a second-degree polynomial function or quadratic function.

For example,

$y = 2x = 2x^1$, $y = \frac{1}{2}x + 1 = \frac{1}{2}x^1 + 1$: linear functions (expressed in terms of x with power 1)

$y = \frac{2}{x} = 2x^{-1}$, $y = x^2 + 1$, $y = x^3 + 2$: not linear functions

> For any constants, $a(\neq 0), b,$
> $ax + b$: expression
> $ax + b = 0$: linear equation
> $ax + b > 0$: linear inequality
> $y = ax + b$: linear function

2. Graphing $y = ax + b$, $a \neq 0$

(1) Graphing $y = ax + b, a \neq 0$ using Translation

The graph of the function $y = ax + b$, $a \neq 0$ is a translation of the graph of the function $y = ax$ with the b units along the y-axis.

> The graph of the function f is the graph of the equation $y = f(x)$.

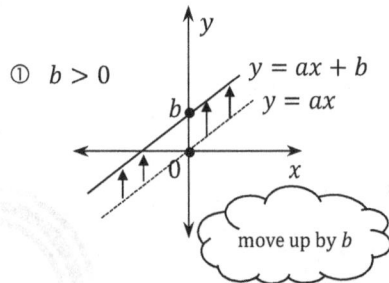

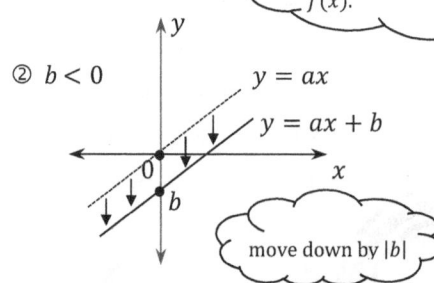

① $b > 0$

$y = ax + b$
$y = ax$

move up by b

② $b < 0$

$y = ax$
$y = ax + b$

move down by $|b|$

(2) Intercepts

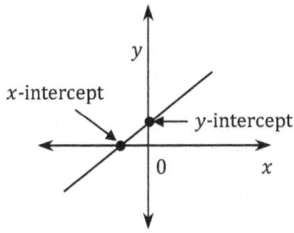

1) The *x-intercept* is the *x*-coordinate of the point where the line crosses (intersects) the *x*-axis.

The *x*-intercept is the value of x when $y = 0$.

Example $y = 2x + 3 \xRightarrow[y=0]{} 0 = 2x + 3 \Rightarrow x = -\dfrac{3}{2}$

$\therefore$ *x*-intercept is $-\dfrac{3}{2}$.

2) The *y-intercept* is the *y*-coordinate of the point where the line crosses (intersects) the *y*-axis.

The *y*-intercept is the value of y when $x = 0$.

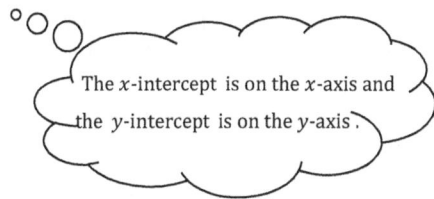

The *x*-intercept is on the *x*-axis and the *y*-intercept is on the *y*-axis.

Example $y = 2x + 3 \xRightarrow[x=0]{} y = 3$

$\therefore$ *y*-intercept is 3.

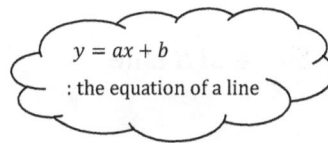

$y = ax + b$
: the equation of a line

(3) Graphing $y = ax + b$, $a \neq 0$ using Intercepts

Steps : 1) Identify the intercepts $\begin{bmatrix} x-\text{intercept: } 0 = ax + b \; ; \; x = -\dfrac{b}{a} \\ y-\text{intercept: } y = a \cdot 0 + b \; ; \; y = b \end{bmatrix}$

2) Graph the intercepts ; Plot the points $\left(-\dfrac{b}{a}, 0\right)$ and $(0, b)$ on a coordinate plane.

3) Connect the points into a straight line.

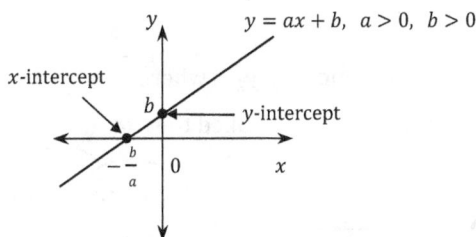

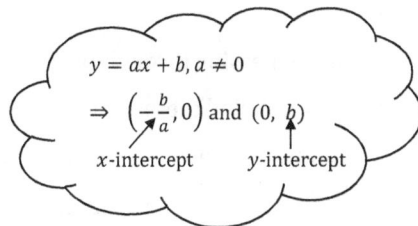

$y = ax + b, a \neq 0$

$\Rightarrow \left(-\dfrac{b}{a}, 0\right)$ and $(0, b)$

$\qquad$ *x*-intercept $\qquad$ *y*-intercept

3. Horizontal and Vertical Lines

(1) A *horizontal line* passes through a point on the y-axis and is parallel to the x-axis.

For any constant k, $y = k$ intersects the y-axis and the value of x doesn't matter at all.

(2) A *vertical line* passes through a point on the x-axis and is parallel to the y-axis.

For any constant k, $x = k$ intersects the x-axis and the value of y doesn't matter at all.

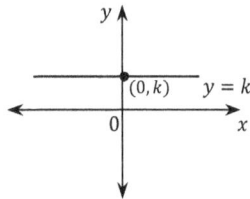

Horizontal line

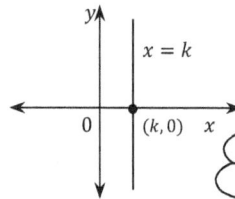

Vertical line

The line $y = 0$ is the x-axis and the line $x = 0$ is the y-axis.

A line passing through (k, p) and (k, q) is $x = k$ ($\because$ the same x-coordinate).

A line passing through (p, k) and (q, k) is $y = k$ ($\because$ the same y-coordinate).

4. The Slope of a Line

To find an equation of a line, consider the slope of the line.

The *slope* of a line measures the steepness and direction of a line. The absolute value of the slope determines a line's steepness. The sign of the slope determines a line's direction.

(1) The Concept of Slope

A line which is not parallel to a coordinate axis may rise from lower left to upper right, or it may fall from upper left to lower right.

For a line passing through the given points (x_1, y_1) and (x_2, y_2), where $x_1 \neq x_2$, the positive number $y_2 - y_1$ is called the *rise*, and the positive number $x_2 - x_1$ is called the *run*.

We define the *slope (m)* of the line by

$$m = \frac{y_2 - y_1}{x_2 - x_1} \quad \text{or} \quad m = \frac{y_1 - y_2}{x_1 - x_2}$$

Subtract the coordinates in the same order in both the numerator and denominator.

Note : $\quad Slope = \dfrac{rise}{run} = \dfrac{change\ in\ y}{change\ in\ x}$

$\quad\quad\quad = Ratio\ of\ the\ vertical\ change\ of\ a\ line\ and\ the\ horizontal\ change\ of\ a\ line.$

① Rise 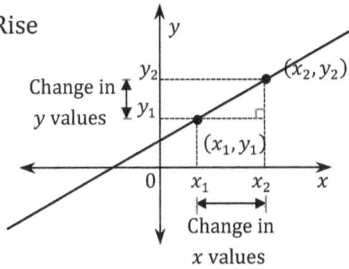 ② Fall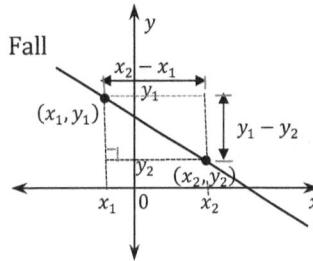

Note 1: For any horizontal line $y = k$, the slope is 0.

$\quad$ ($\because$ All the points (coordinates) on the line $y = k$ have same y value, k. That is, there is no change in y.

$\quad$ So, $slope = \dfrac{0}{change\ in\ x} = 0$)

Note 2: For any vertical line $x = k$, the slope does not exist.

$\quad$ ($\because$ All the points (coordinates) on the line $x = k$ have same x value, k. That is, there is no change in x.

$\quad$ So, $slope = \dfrac{change\ in\ y}{0}$.

$\quad$ This equation is undefined because a denominator can't be divided by zero.

$\quad$ Therefore, vertical lines have no slope (slope is undefined).)

Horizontal lines have 0 (zero) slope.

Vertical lines have no slope.

(lying : "zero" steepness)

(standing up : "no" steepness)

The magnitude of the slope, either positive or negative is a measure of steepness for the line.

(2) Properties of Slope

1) For two lines $y = ax + b,\ a \neq 0$ and $y = cx + d,\ c \neq 0$,

the line $y = ax + b$ is steeper than the line $y = cx + d$ if $|a| > |c|$

The larger the absolute value of the slope, the steeper the line will be.

Example

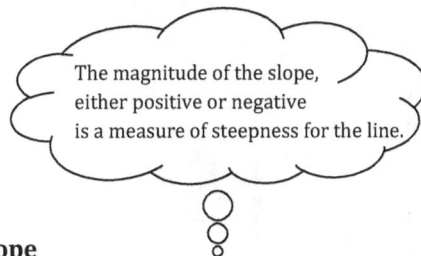 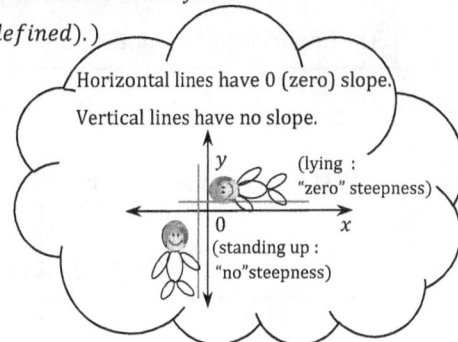

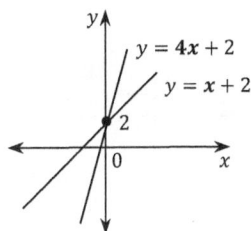 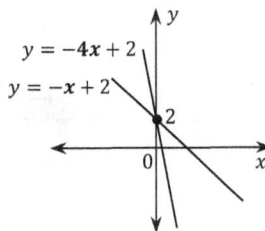 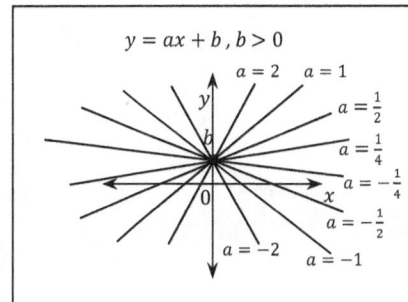

2) For a linear function $y = ax + b$, $a \neq 0$, the slope is a.

① $a > 0$ (positive slope) : The line goes uphill from left to right (the line rises to the right.)

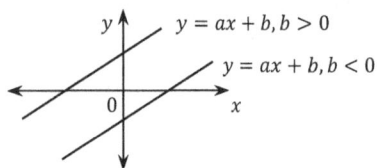

x values increase $\Rightarrow$ y values increase

② $a < 0$ (negative slope) : The line goes downhill from left to right (the line falls to the right.)

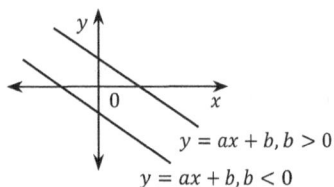

x values increase $\Rightarrow$ y values decrease

(3) Graphing $y = ax + b$, $a \neq 0$ using slope and y-intercept

1) Plot the y-intercept on a coordinate plane.

2) Using the slope, find the other point.

3) Connect the two points into a straight line.

$Note:$ $y = ax + b$ $a > 0$, $b > 0$

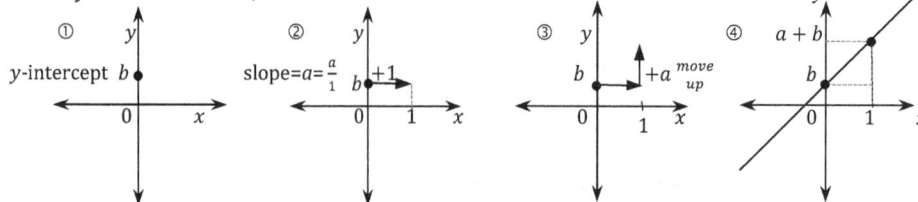

Example

$$y = -\frac{5}{2}x + 1$$

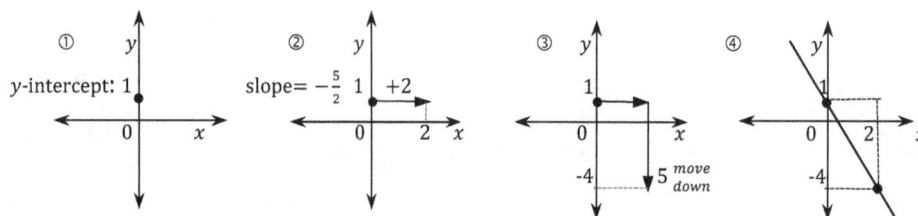

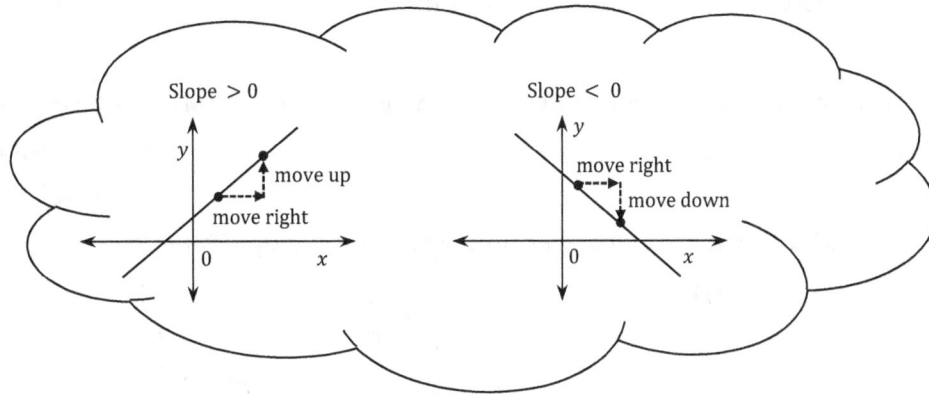

2-2 Lines and their Equations

For a given linear equation $ax + by + c = 0$ (constants $a \neq 0$, $b \neq 0$, and c),
the *linear function* is represented by

$$y = -\frac{a}{b}x - \frac{c}{b} \quad (\text{constants } a \neq 0, \ b \neq 0, \text{ and } c).$$

The graphs of the linear equation and the linear function are the same lines.

To find a slope and y-intercept,
transform the standard form, $ax + by + c = 0$
into the form of $y = mx + P$.

Note 1 : The standard form of an equation for a line is

$$ax + by + c = 0, \ a \neq 0 \ or \ b \neq 0.$$

However, $y = ax + b$ is not the standard form of a line.
(∵ If a (x-coefficient) $= 0$, then $y = b$ (: a horizontal line).
However, vertical line $x = k$ can't be obtained
because the y-coefficient is $1(\neq 0)$, that is, the y- term does not disappear.)

Note 2 : Graphing $ax + by + c = 0$:

   ① $a \neq 0, \ b \neq 0 \Rightarrow y = -\frac{a}{b}x - \frac{c}{b}$: *line with slope* $= -\frac{a}{b}$ *and y-intercept* $= -\frac{c}{b}$

   ② $a \neq 0, \ b = 0 \Rightarrow ax + c = 0$; $x = -\frac{c}{a}$: *vertical line parallel to the y-axis.*

   ③ $a = 0, \ b \neq 0 \Rightarrow by + c = 0$; $y = -\frac{c}{b}$: *horizontal line parallel to the x-axis.*

1. Methods of Creating the Equations of Lines

(1) A line which has slope m and y-intercept b (*Slope-Intercept Form*)

$\Rightarrow\ y = mx + b$

x-coefficient is slope m.

(2) A line which has slope m and passes through a point (x_1, y_1) (*Point-Slope Form*)

$\Rightarrow\ y - y_1 = m(x - x_1)$

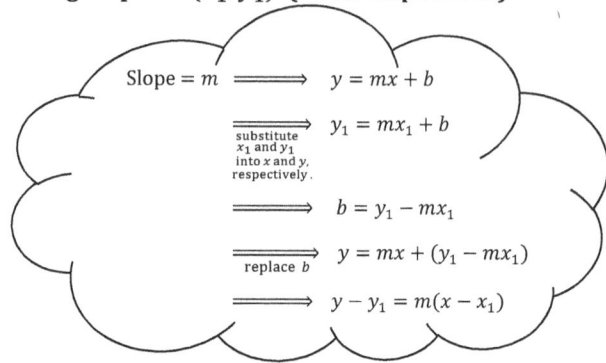

Slope $= m$ $\implies$ $y = mx + b$

$\xrightarrow[\substack{x_1 \text{ and } y_1 \\ \text{into } x \text{ and } y, \\ \text{respectively}.}]{\text{substitute}}$ $y_1 = mx_1 + b$

$\implies$ $b = y_1 - mx_1$

$\xrightarrow{\text{replace } b}$ $y = mx + (y_1 - mx_1)$

$\implies$ $y - y_1 = m(x - x_1)$

(3) A line which passes through two different points (x_1, y_1) and (x_2, y_2), $x_1 \neq x_2$

$\Rightarrow\ ①$ Find the slope m :

$$m = \frac{y_2 - y_1}{x_2 - x_1}\ \text{ or }\ m = \frac{y_1 - y_2}{x_1 - x_2}$$

② Use the slope m and one of two points. Using point-slope form,

$$y - y_1 = m(x - x_1)\ \text{ or }\ \ y - y_2 = m(x - x_2)$$

(4) A line which has an x-intercept p and y-intercept q : the same as a line which passes through two points $(p, 0)$ and $(0, q)$

$\Rightarrow\ ①$ Find the slope m :

$$m = \frac{q - 0}{0 - p}\ \text{ or }\ m = \frac{0 - q}{p - 0}$$

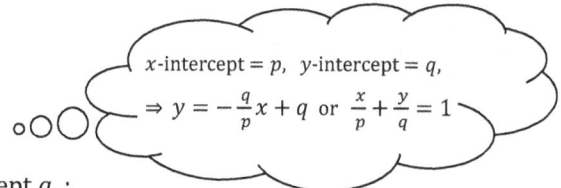

x-intercept $= p$, y-intercept $= q$,

$\Rightarrow y = -\dfrac{q}{p}x + q$ or $\dfrac{x}{p} + \dfrac{y}{q} = 1$

② Use the line which has slope m and y-intercept q :

$$y = mx + q$$

2. Equations of Lines

> **(1) Vertical line** : $x = k$
>
> **(2) Horizontal line** : $y = k$
>
> **(3) General linear equation** : $ax + by + c = 0$
>
> **(4) Slope-intercept form** : $y = mx + b$
>
> **(5) Point-slope form** : $y - y_1 = m(x - x_1)$

> Parallel lines (⟋⟋) never intersect even when extended.
> Perpendicular lines (←□→) intersect at one point and form a right angle.

3. Parallel and Perpendicular Lines

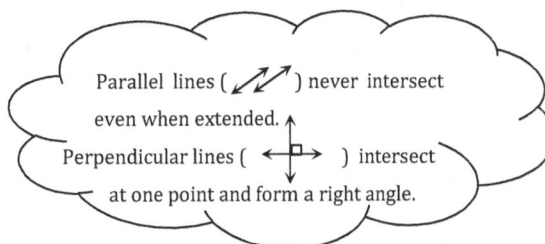

For any two different lines $y = m_1 x + a$ and $y = m_2 x + b$, parallel and perpendicular lines are identified by comparing their slopes, m_1 and m_2.

---

(1) $m_1 = m_2$: **The slopes are the same.**

⇒ **parallel lines**

(2) $m_1 \cdot m_2 = -1 \left(\text{or } m_1 = -\dfrac{1}{m_2} \; ; \; m_2 = -\dfrac{1}{m_1} \right)$: **Slopes are negative reciprocals of each other.**

⇒ **perpendicular lines**

Note 1: *For a line i which has the equation $y = \frac{2}{3}x + 2$, the slope of the line i is $\frac{2}{3}$.*

So, the parallel line j has the same slope of $\frac{2}{3}$, and the slope of a perpendicular line k is $-\frac{3}{2}$.

Note 2: ① *Lines i and j are parallel.*

⇒ *These two lines have the same slope and different y-intercepts.*

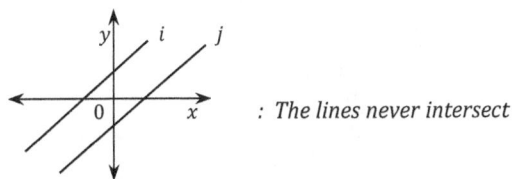

: The lines never intersect

② *Lines i and k are perpendicular.*

⇒ *Their slopes are negative reciprocals of each other.*

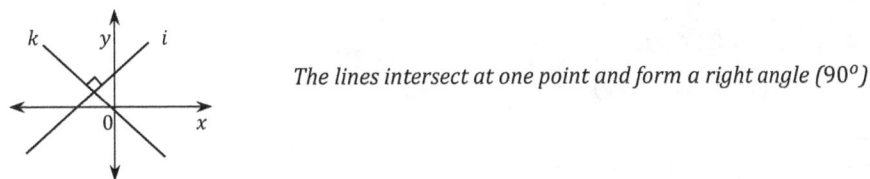

The lines intersect at one point and form a right angle (90°)

③ *Lines i and l coincide.*

⇒ *These two lines have the same slope and the same y-intercept.*

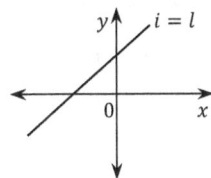

Note 3 : ① *These lines are symmetrical with respect to the x-axis.*

This happens whenever (x, y) *and* $(x, -y)$ *are on the same graph.*

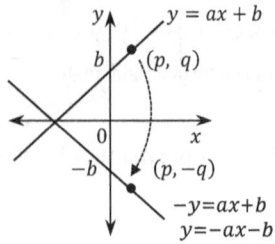

② *These lines are symmetrical with respect to the y-axis.*

This happens whenever (x, y) *and* $(-x, y)$ *are on the same graph.*

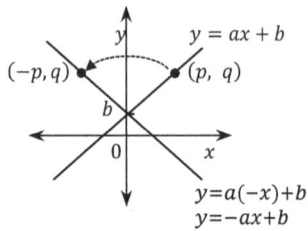

4. Translation and Symmetry for $y = ax$, $a \neq 0$

(1) $(x, y) \longrightarrow (x + m, \ y + n)$: Translate m units along the x-axis and n units along the y-axis

$\Rightarrow y - n = a(x - m)$

(2) Symmetry

1) along the x-axis : $(x, y) \longrightarrow (x, -y)$

2) along the y-axis : $(x, y) \longrightarrow (-x, y)$

3) along the origin $(0,0)$: $(x, y) \longrightarrow (-x, -y)$

4) When $y = x$: $(x, y) \longrightarrow (y, x)$

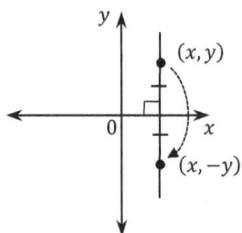

Figure 1)

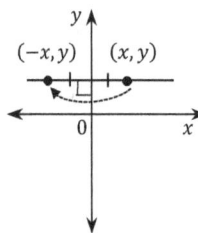

Figure 2)

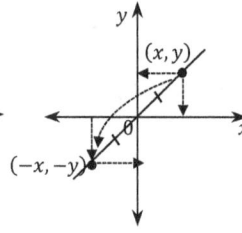

Figure 3)

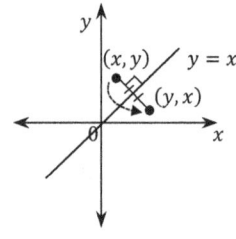

Figure 4)

2-3 Solving Equations by Graphing

$$ax + by + c = 0, \ a \neq 0 \text{ or } b \neq 0$$
(Linear equation)

$$\xrightarrow[\text{solve for } y]{} \quad y = -\frac{a}{b}x - \frac{c}{b} \ (\text{ linear function })$$

For any constants $a, b,$ and $c,$

$$ax + by + c = 0, \ a \neq 0 \text{ or } b \neq 0$$

is the standard form of equation for a straight line.

The graph of an equation with variables x and y consists of points in the coordinate plane whose coordinates (x, y) make the equation true.

The relationship of the two linear equations $\begin{cases} ax + by + c = 0 \\ a'x + b'y + c' = 0 \end{cases}$ in a coordinate plane is one of the following cases :

(1) Case 1 Parallel

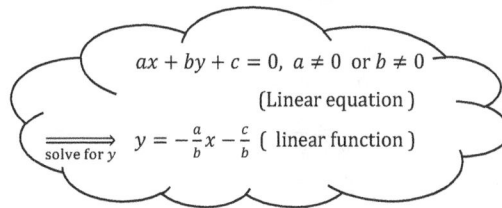

$\Rightarrow$ ① no intersection

② no solution for the system

③ same slopes $\left(-\dfrac{a}{b} = -\dfrac{a'}{b'} \ ; \dfrac{a}{b} = \dfrac{a'}{b'} \ ; \ ab' = a'b \ ; \dfrac{a}{a'} = \dfrac{b}{b'} \right)$

④ different y-intercepts $\left(-\dfrac{c}{b} \neq -\dfrac{c'}{b'} \ ; \dfrac{c}{b} \neq \dfrac{c'}{b'} \ ; \ b'c \neq bc' \ ; \dfrac{b}{b'} \neq \dfrac{c}{c'} \right)$

(2) Case 2 Intersecting at one point (perpendicular or not)

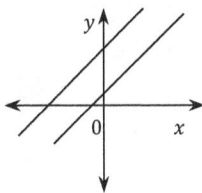

$\Rightarrow$ ① 1 intersection

② 1 solution for the system

③ different slopes $\left(-\dfrac{a}{b} \neq -\dfrac{a'}{b'} \right)$

(3) Case 3 Coinciding

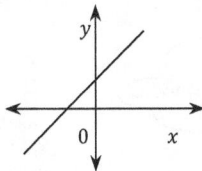

$\Rightarrow$ ① unlimited number of intersections

② unlimited number of solutions for the system

③ the same slopes $\left(-\dfrac{a}{b} = -\dfrac{a'}{b'} \right)$

④ the same y-intercepts $\left(-\dfrac{c}{b} = -\dfrac{c'}{b'} \right)$

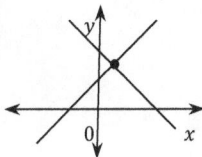

1 intersection

$ax + by + c = 0$

$(\alpha, \beta) \longrightarrow$ 1 solution

The solution $(x = \alpha, y = \beta)$ of two equations (a system of equations) with two variables is the intersection point (α, β) between the two lines.

Note: $\begin{cases} ax + by + c = 0 \\ a'x + b'y + c' = 0 \end{cases}$ $\xrightarrow[\text{solve for } y]{}$ $\begin{cases} y = mx + p \\ y = m'x + p' \end{cases}$

① $\dfrac{a}{a\prime} = \dfrac{b}{b\prime} \neq \dfrac{c}{c\prime}$ $\Rightarrow$ *parallel* $\Leftarrow$ $m = m'$ *and* $p \neq p'$

② $\dfrac{a}{a\prime} \neq \dfrac{b}{b\prime}$ $\Rightarrow$ *intersecting at one point* $\Leftarrow$ $m \neq m'$

③ $\dfrac{a}{a\prime} = \dfrac{b}{b\prime} = \dfrac{c}{c\prime}$ $\Rightarrow$ *coinciding* $\Leftarrow$ $m = m'$ *and* $p = p'$

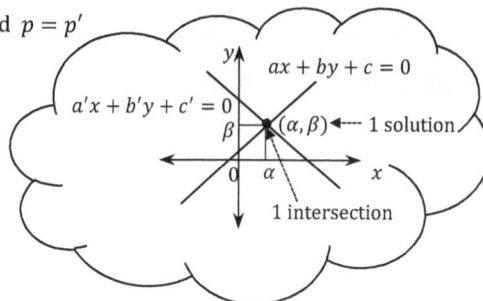

2-4 Graphing with Absolute Values

Note : $|a| = \begin{cases} a, & a \geq 0 \\ -a, & a < 0 \end{cases}$

For $y = |x|$, all real numbers have exactly one absolute value.

So, $y = |x|$ is a function.

1. Graphing of an absolute value function, $y = |x|$.

Case 1: $x \geq 0 \Rightarrow y = x$ Case 2: $x < 0 \Rightarrow y = -x$

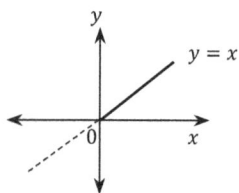

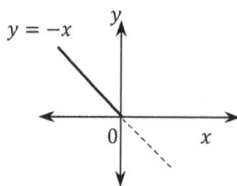

Therefore, the graph of $y = |x|$ is

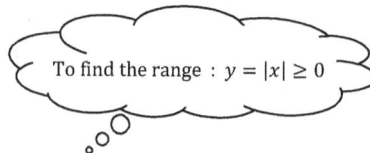

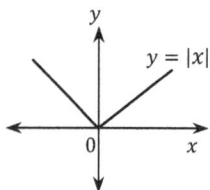

To find the range : $y = |x| \geq 0$

where, domain=all real numbers and

range= $\{ y|\, y \geq 0 \}$ all nonnegative real numbers.

2. Graphing $y = |x + 2| - 1$

To get the range ;
Since $|x + 2| \geq 0$,
$y = |x + 2| - 1 \geq -1$

Case 1 : $x + 2 \geq 0$ ($x \geq -2$)
$\Rightarrow y = x + 1$

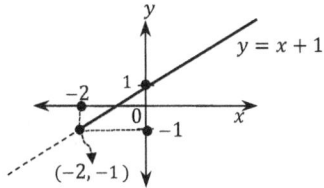

Case 2 : $x + 2 < 0$ ($x < -2$)
$\Rightarrow y = -x - 3$

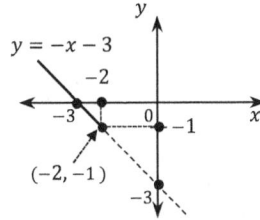

If $x + 2 < 0$, then
$y = |x + 2| - 1$
$= -(x + 2) - 1$
$= -x - 3$

Therefore, the graph of $y = |x + 2| - 1$ is

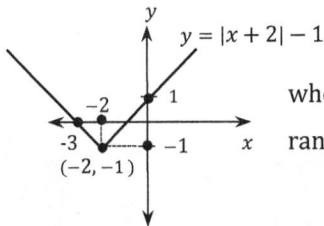

where, domain = all real numbers and

range = $\{ y | y \geq -1 \}$ all real numbers greater than or equal to -1.

Exercises

#1 Identify linear functions. Mark o for a linear function or × for a non-linear function.

(1) $y = 2 + x$

(2) $2x + y = 3$

(3) $y = \frac{2}{x} + 1$

(4) $x^2 + y^2 = 1$

(5) $y = x^2 + 2x + 1$

(6) $x + y = 0$

(7) $y = 2(x + 1) - 2x$

(8) $y = x^2 - (x + 2)^2$

(9) $y = 1$

(10) $xy = 2$

#2 Find the following values for the linear function $f(x) = 2x + 3$:

(1) $f(0)$

(2) $f(1) + f(-1)$

(3) $\frac{1}{2}f(-2) \cdot f(\frac{1}{2})$

(4) $f(f(2))$

#3 Find the following values for the given linear functions with a condition :

(1) $f(1)$ for $f(x) = 2ax + 1$ with $f(-1) = 3$

(2) $\frac{a}{2}$ for $f(x) = \frac{1}{2}x + 5$ with $f\left(\frac{a}{2}\right) = -a$

(3) $a + b$ for $f(x) = 3ax - 2$ with $f(-1) = 4$ and $f(b) = 1$

(4) $a + \dfrac{1}{a}$ when $f(x) = 3x - 1$ passes through the point $(a, a + 3)$.

(5) $a - b$ when $f(x) = ax + 2$ passes through both point $(1, 3)$ and point $(2, b)$.

#4 Find the value of $a + b$ for which

(1) The graph of $y = ax + 2$ is translated by b along the y-axis from a graph of $y = 3x - 5$.

(2) The graph is translated by a along the y-axis from a graph of $y = 2x + 4$ and passes through both point $(a + 1, -2)$ and point $\left(-\dfrac{1}{3}, b\right)$.

(3) A point $(-1, 1)$ is on the graph of $y = -2x + a$. If the graph is translated by b along the y-axis, then it will pass through the point $(3, -4)$.

#5 Find the x-intercept and y-intercept.

(1) The linear function $y = ax + b$ passes through both point $(1, 2)$ and point $(-1, 4)$.

(2) The graph of $y = ax + b$ intersects the graph of $y = 2x + 3$ on the x-axis. It also intersects the graph of $y = -5x - 6$ on the y-axis.

(3) The area surrounded by the graph of $y = \frac{1}{2}x + a$ $(a > 0)$, the x-axis, and the y-axis is 36.

#6 Find the area of the polygon surrounded by

(1) A graph of $y = -\frac{2}{3}x + 2$, the x-axis, and the y-axis

(2) The graphs of $y = x + 4$ and $y = -\frac{1}{2}x + 4$ and the x-axis

(3) The graphs of $y = \frac{1}{3}x + 3$ and $y = \frac{1}{9}x + 1$ and the y-axis

(4) The graphs of $x = 3$ and $y = 4$, the x-axis, and the y-axis

---

#7 Find the slopes of the lines containing the given two points.

(1) $(2, 3)$ and $(4, -1)$

(2) $(0, 5)$ and $(-5, 0)$

(3) $(4, -1)$ and $(2, -5)$

(4) $(-3, 2)$ and $(0, -1)$

(5) $(-2a, 0)$ and $(0, -4a), a \neq 0$

#8 Find the slope m and y-intercept b of each line.

(1) $2x + 3y = 4$

(2) $3y = 4x - 5$

(3) $x + 3y = -2$

(4) $y + 5 = 3x$

(5) $y = 2x$

(6) $y - 2 = 0$

#9 Find an equation in the standard form for each line.

(1) with y-intercept -3 and slope 2

(2) with y-intercept 5 and slope 0

(3) with x-intercept 5 and slope $-\frac{2}{3}$

(4) with x-intercept -3 and slope -2

(5) through $(1, 2)$ with slope 3

(6) through $(3, -4)$ with slope -2

(7) through $(2, 3)$ with undefined slope

(8) through $(-2, 3)$ with y-intercept -1

(9) through $(2, 4)$ with x-intercept -5

(10) through $(3, 1)$ and $(-2, 4)$

(11) through $(-2, -3)$ and $(-1, 5)$

(12) with x-intercept -3 and y-intercept 3

(13) with x-intercept $\frac{3}{2}$ and y-intercept -4

(14) Vertical line through $(-1, 2)$

(15) Horizontal line through $(3, -4)$

#10 Find an equation for the line through $(2, 3)$ which is :

(1) parallel to the line $y = 2x - 5$

(2) parallel to the line $y = -3x + 1$

(3) parallel to the line $x = 4$

(4) parallel to the line $y = -2$

(5) parallel to the line $3x + 4y = 5$

(6) perpendicular to the line $y = \frac{2}{3}x - 1$

(7) perpendicular to the line $x + 3y = -3$

(8) perpendicular to the line $x = 5$

(9) perpendicular to the line $y = -2$

#11 Find the value of a for the following lines

(1) through $(2, 3)$ and $(1, -a)$ with slope 2

(2) through $(2a - 1, -2)$ and $(-1, 1)$ with slope -2

(3) through $(1, -2)$, $(-3, 2)$, and $(-a + 1, -5)$

(4) through $(2a + 1, -4)$, $(2, 5)$, and $(2, -3)$

(5) through $(-3, 3)$, $(3, a - 1)$, and $(0, 3)$

(6) through $(a, 2a - 3)$ and $(-a - 1, 3 + 4a)$ and parallel to x-axis

(7) through $(-3a + 1, -5)$ and $(2a - 1, a + 3)$ and perpendicular to x-axis

(8) through $(3, -2a)$ and $(2a - 1, -3a + 2)$ and parallel to the y-axis

(9) through $(-1, 5)$ and $(2, -4)$ and parallel to the line $ax + 3y + 5 = 0$

#12 Find the value of a such that the line $ax + 2y = 5$

    (1) is parallel to the line $2x + 3y = -2$.

    (2) is perpendicular to the line $y = -2x + 3$.

    (3) coincides with the line $6y = -4x + 15$.

#13 Find the value of ab for which

    (1) the system $\begin{cases} x - 3y = a \\ 2x + by = 3 \end{cases}$ has the intersection point $(2, 3)$.

    (2) the system $\begin{cases} -ax + by = 4 \\ 2ax + 3by = 2 \end{cases}$ has the intersection point $(-1, 2)$.

    (3) the system $\begin{cases} px + y = 3 \\ 2x - 3y = q \end{cases}$ has no intersection when $p = a$, $q \neq b$.

(4) the system $\begin{cases} 2ax + 4y = -3 \\ 3x + 6y = 2b \end{cases}$ has an unlimited number of intersections.

#14 Find the value of a such that

(1) the system $\begin{cases} ax + y = -2 \\ -3x + 2y = 4 \end{cases}$ has no solution.

(2) the system $\begin{cases} 2x - ay + 3 = 0 \\ x + 3y - 2 = 0 \\ 2x + y + 1 = 0 \end{cases}$ has one solution.

(3) the system $\begin{cases} x - 3y = 2 \\ 2x + y = -3 \end{cases}$ has a solution $(2a, -1)$.

(4) the line $2ax + 3y - 1 = 0$ passes through the intersection of the system $\begin{cases} x - 2y = 3 \\ 2x + 2y = 1 \end{cases}$.

#15 Find the equation of each line such that

(1) the line passes through the intersection of the system $\begin{cases} x + 2y = 3 \\ 3x + y = -2 \end{cases}$

and runs parallel to the y-axis.

(2) the line passes through the intersection of the system $\begin{cases} -x + y + 2 = 0 \\ 2x + y - 3 = 0 \end{cases}$

and runs perpendicular to the x-axis.

(3) the line passes through the intersection of the system $\begin{cases} 2x - y + 3 = 0 \\ x + 2y + 4 = 0 \end{cases}$

and runs parallel to the line $3x + 2y = 5$.

#16 Find the area of the polygon surrounded by the two lines ($3x + 4y - 16 = 0$ and $3x - 2y - 10 = 0$), the x-axis, and the y-axis.

#17 The area of a polygon surrounded by $y = x$, $y = ax + b$ ($b > 0$) which has x-intercept 6, and the x-axis is 12. Find the area of a polygon surrounded by those two lines and the y-axis.

#18 The perimeter of a rectangle with the length 5 inches and the width x inches is y square inches. Find the relationship between x and y.

#19 Richard drives a car 15 miles total, from place A to place B. He begins at a speed of 30 miles per hour. x minutes after departing, he has y miles more to go to arrive at place B. Find the relationship between x and y.

#20 Richard and Nichole drive toward each other from opposite starting points 4 miles apart. Richard drives at a speed of 40 miles per hour and Nichole drives at a speed of 35 miles per hour. After x minutes, the distance between the two is y miles. Find the relationship between x and y.

#21 Graph the following lines

(1) $y = -|x|$

(2) $y = -|x + 3| + 2$

(3) $y = |x| + x$

(4) $|y - 1| = x + 2$

Algebra

Part III
Functions

Chapter 3

Quadratic Equations
Level III

Chapter 3 Quadratic Functions

3-1 Quadratic Functions and their Graphs

1. **Quadratic Functions**
   (1) Definition
   (2) Parabola
2. **Graph of Quadratic Functions**
   (1) Graphing $y = ax^2$
   (2) Graphing $y = ax^2 + q$
   (3) Graphing $y = a(x - p)^2$
   (4) Graphing $y = a(x - p)^2 + q$

3-2 Properties of Quadratic Functions

1. Graphing $y = ax^2 + bx + c$
2. Intercepts
   (1) x-intercept
   (2) y-intercept
3. Translation and Symmetry
   (1) $y = ax^2$
      1) Translation
      2) Symmetry
   (2) $y = a(x - p)^2 + q$
      1) Translation
      2) Symmetry

4. **Conditions for the y-value**
   (1) $a > 0$
   (2) $a < 0$
5. **Properties of $y = a(x - p)^2 + q$**
   (1) The signs of the x^2-coefficient
   (2) The sides of the axis of Symmetry
   (3) The signs of the y-intercept

3-3 Solving Quadratic Functions

1. Equations of Quadratic Functions
2. Maximum or Minimum Values of Quadratic Functions
   (1) $y = a(x - p)^2 + q$
   (2) $y = ax^2 + bx + c$
   (3) When a Maximum Value or Minimum Value is given:
3. Steps for Solving Word Problems

CHAPTER 3

Chapter 3. Quadratic Functions

3-1 Quadratic Functions and their Graphs

1. Quadratic Functions

> $ax^2 + bx + c = 0,\ a \neq 0$: quadratic equation
>
> $y = ax^2 + bx + c,\ a \neq 0$: quadratic function

(1) Definition

For any function $y = f(x)$, a function represented by the form

$$y = ax^2 + bx + c, \quad \text{with constants } a \neq 0,\ b,\ c$$

is called a *quadratic function*.

> Quadratic
>
> : The highest power in a polynomial is 2.

Example $y = 3x^2,\ y = -x^2 + 1,\ y = 2x^2 + 3x,\ y = \frac{1}{2}x^2 - x + 2$

> the graph of a linear function = line
>
> the graph of a quadratic function = parabola

(2) Parabola

1) A graph of a quadratic function $y = ax^2 + bx + c, a \neq 0$ on a coordinate plane
    is called a *parabola*.

2) The symmetrical axis (central line) of a parabola is called the *axis* of the parabola or *the axis of symmetry*. A parabola is symmetrical with respect to the axis of symmetry.

3) The intersection point of the parabola and the axis of symmetry is called the *vertex*.

4) There are two types of parabola :

①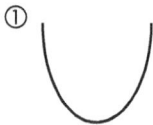

opens upward

The vertex is the minimum point.

(lowest point)

②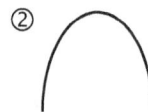

opens downward

The vertex is the maximum point.

(highest point)

Note : On the graph of the simplest quadratic equation $y = x^2$, the equation of the axis of symmetry is the line $x = 0$ (y axis) and the coordinates of the vertex are $(0,0)$.

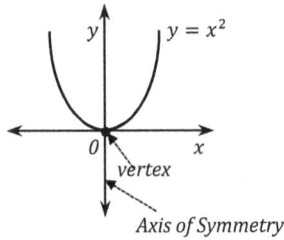

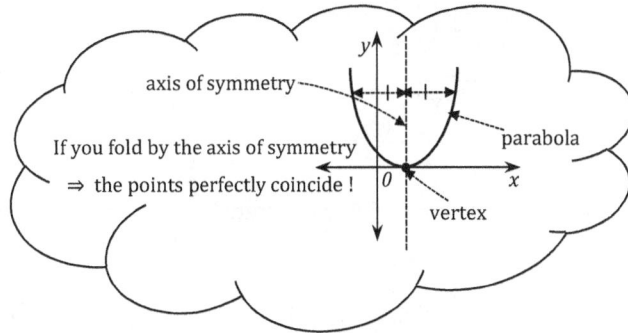

2. Graphing of Quadratic Functions

(1) Graphing $y = ax^2$

1) $a > 0$

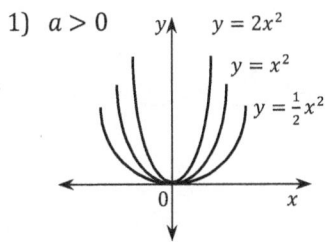

2) $a < 0$

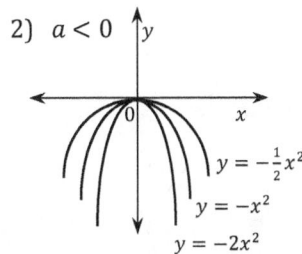

① Vertex : $(0, 0)$

② Axis of symmetry : $x = 0$ (y-axis)

③ $a > 0 \Rightarrow$ open upward and $y = f(x) \geq 0$

    $a < 0 \Rightarrow$ open downward and $y = f(x) \leq 0$

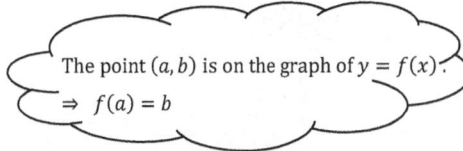

The point (a, b) is on the graph of $y = f(x)$.
$\Rightarrow f(a) = b$

④ The larger the magnitude of the absolute value of a, the narrower the width of the parabola.

Note : The graphs of $y = ax^2$ and $y = -ax^2$ are symmetric along the x-axis.

When $a > 0$,	When $a < 0$,
$x \uparrow \Rightarrow f(x) \uparrow$	$x \uparrow \Rightarrow f(x) \downarrow$
(If the value of x increases, then the value of $f(x)$ increases.)	(If the value of x increases, then the value of $f(x)$ decreases.)

(2) Graphing $y = ax^2 + q$, $a \neq 0$

1) $a > 0$, $q > 0$

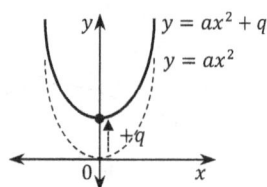

2) $a < 0$, $q > 0$

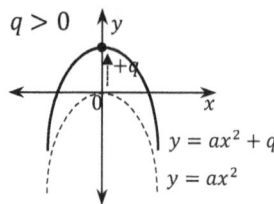

The graph of $y = ax^2 + q$ is a translation of the graph of $y = ax^2$, with the q units along the y-axis.

① Vertex $(0, q)$

② Axis of symmetry $x = 0$ (y-axis)

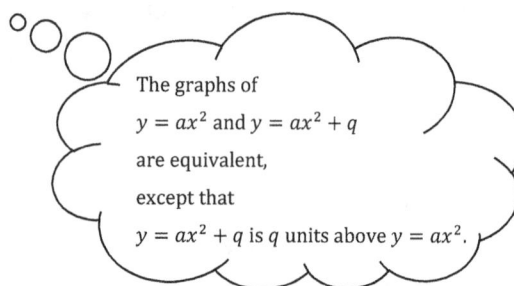

The graphs of
$y = ax^2$ and $y = ax^2 + q$
are equivalent,
except that
$y = ax^2 + q$ is q units above $y = ax^2$.

(3) Graphing $y = a(x - p)^2$, $a \neq 0$

1) $a > 0$, $p > 0$

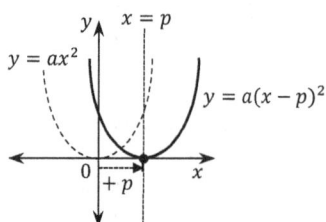

2) $a < 0$, $p > 0$

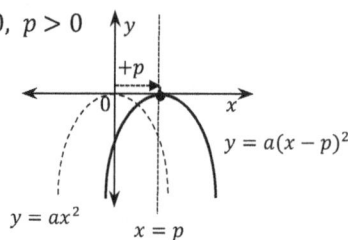

The graph of $y = a(x - p)^2$ is a translation of the graph of $y = ax^2$,

with the p units along the x-axis.

① Vertex $(p, 0)$

② Axis of symmetry $x = p$

(4) Graph of $y = a(x - p)^2 + q$, $a \neq 0$

> Shift the graph of $y = ax^2$
> p units to the right and q units up.

1) $a > 0$, $p > 0$, $q > 0$

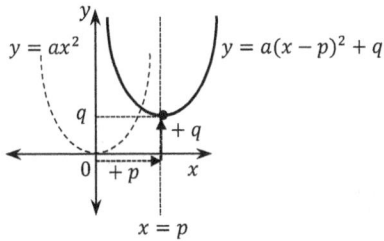

2) $a < 0$, $p > 0$, $q > 0$

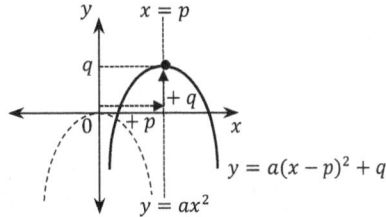

The graph of $y = a(x - p)^2 + q$ is a translation of the graph of $y = ax^2$, with the p units along the x-axis and the q units along the y-axis.

① Vertex (p, q)

② Axis of symmetry $x = p$

Note : Graphing $y = a(x + p)^2 - q$, $a \neq 0$

$a < 0$, $p > 0$, $q > 0$

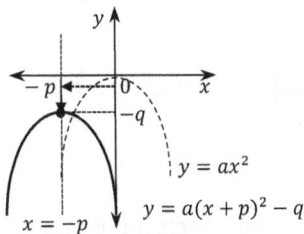

> A parabola with vertex (m, n)
> $\Leftrightarrow y = a(x - m)^2 + n$

The graph of $y = a(x + p)^2 - q$ is a translation of the graph of $y = ax^2$, with the $-p$ units along the x-axis and the $-q$ units along the y-axis.

① *Vertex $(-p, -q)$*

② *Axis of symmetry $x = -p$*

3-2 Properties of Quadratic Functions

1. Graphing $y = ax^2 + bx + c$

To find the vertex and the axis of symmetry for a graph, transform the graph of the general quadratic function form $y = ax^2 + bx + c$, $a \neq 0$ to the graph of a standard form :

$$y = a(x - p)^2 + q, \ a \neq 0$$

Note : $y = ax^2 + bx + c = a\left(x^2 + \frac{b}{a}x\right) + c = a\left(\left(x + \frac{b}{2a}\right)^2 - \left(\frac{b}{2a}\right)^2\right) + c$

$\qquad\quad = a\left(x + \frac{b}{2a}\right)^2 - \frac{b^2}{4a} + c = a\left(x + \frac{b}{2a}\right)^2 - \frac{b^2 - 4ac}{4a}$

(1) $y = ax^2 + bx + c \xrightarrow[\text{transform}]{} y = a\left(x + \frac{b}{2a}\right)^2 - \frac{b^2 - 4ac}{4a}$

(2) Vertex : $\left(-\dfrac{b}{2a}, -\dfrac{b^2 - 4ac}{4a}\right)$

(3) Axis of symmetry : $x = -\dfrac{b}{2a}$

2. Intercepts

(1) *x*-intercept

The *x-intercept* is the x-coordinate of the point where the graph intersects the x-axis.

x-intercept is the value of x when $y = 0$.

(2) *y*-intercept

The *y-intercept* is the y-coordinate of the point where the graph intersects the y-axis.

y-intercept is the value of y when $x = 0$.

Note 1

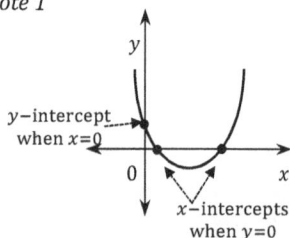

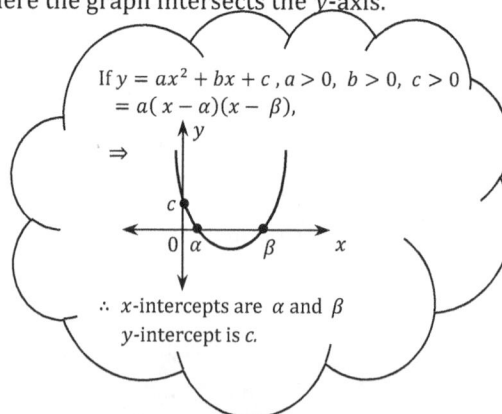

If $y = ax^2 + bx + c$, $a > 0$, $b > 0$, $c > 0$
$= a(x - \alpha)(x - \beta)$,

$\Rightarrow$

$\therefore$ x-intercepts are α and β
y-intercept is c.

Note 2

(1) *If* $y = ax^2 + bx + c = a(x - \alpha)(x - \beta)$,

    the equation $ax^2 + bx + c = 0$ *has two different solutions.*

      (using $D = b^2 - 4ac > 0$ *)*

    $\Rightarrow$ *The parabola has 2 different x-intercepts.*

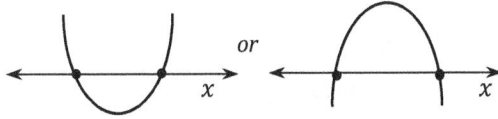

(2) *If* $y = ax^2 + bx + c = a(x - \alpha)^2$,

    the equation $ax^2 + bx + c = 0$ *has one solution (the double root).*

      (using $D = b^2 - 4ac = 0$ *)*

    $\Rightarrow$ *The parabola has only 1 x-intercept.*

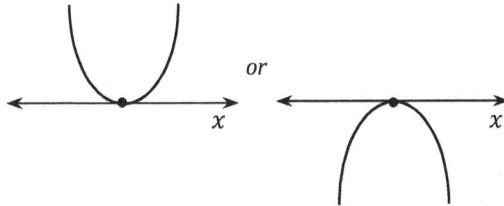

(3) *If* $y = ax^2 + bx + c$,

    the equation $ax^2 + bx + c = 0$ *has no solution.*

      (using $D = b^2 - 4ac < 0$ *)*

    $\Rightarrow$ *The parabola has no x-intercept.*

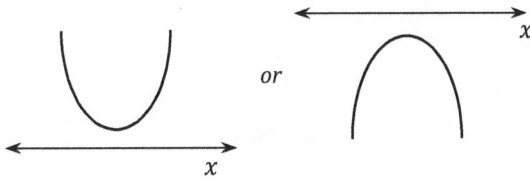

3. Translation and Symmetry

(1) $y = ax^2$

1) Translation

To find the translation of the graph of $y = ax^2$,

with the p units along the x-axis and the q units along the y-axis.

$\Rightarrow$ Substitute $x - p$ into x and substitute $y - q$ into y.

$\Rightarrow$ $y - q = a(x - p)^2$

$\Rightarrow$ $y = a(x - p)^2 + q$

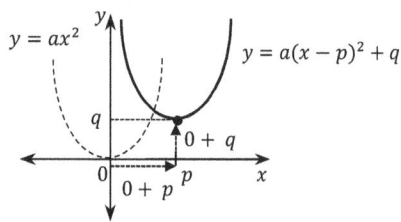

$y = ax^2$

$\Rightarrow$ $y = a(x - p)^2 + q$

2) Symmetry

① To create symmetry along the x-axis, substitute $-y$ into y.

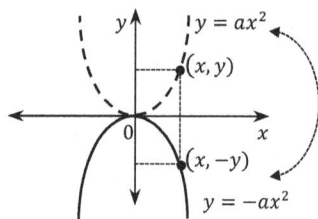

$y = ax^2$

$\Rightarrow$ $-y = ax^2$

$\Rightarrow$ $y = -ax^2$

② To create symmetry along the y-axis, substitute $-x$ into x.

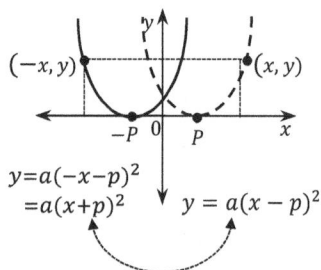

$y = a(x - p)^2$

$\Rightarrow$ $y = a(-x - p)^2$

$\Rightarrow$ $y = a(x + p)^2$

(2) $y = a(x - p)^2 + q$

1) Translation

To find the translation of the graph of $y = a(x - p)^2 + q$,

with the m units along the x-axis and the n units along the y-axis

$\Rightarrow$ Substitute $x - m$ into x and $y - n$ into y.

$\Rightarrow y - n = a(x - m - p)^2 + q$

$\Rightarrow y = a(x - (p + m))^2 + q + n$

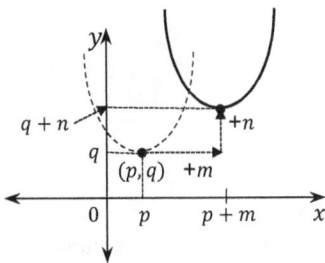

$y = a(x - p)^2 + q$

$\Rightarrow y = a(x - (p + m))^2 + (q + n)$

2) Symmetry

① To create symmetry along the x-axis, substitute $-y$ into y.

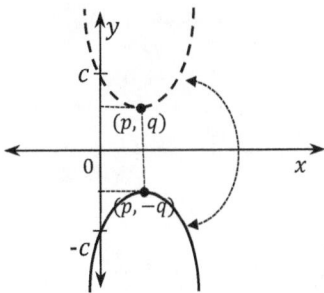

$y = a(x - p)^2 + q$

$\Rightarrow -y = a(x - p)^2 + q$

$\Rightarrow y = -a(x - p)^2 - q$

② To create symmetry along the y-axis, substitute $-x$ into x.

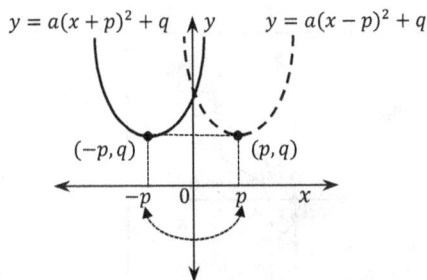

$y = a(x - p)^2 + q$

$\Rightarrow y = a(-x - p)^2 + q$

$\Rightarrow y = a(x + p)^2 + q$

4. Conditions for the y-value

$$y = ax^2 + bx + c = a(x - p)^2 + q$$

(1) $a > 0$

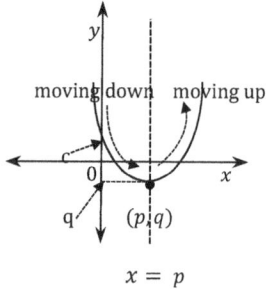

$x = p$

① When $x < p$,

If the x–values are increasing. $\Rightarrow$ the y–values are decreasing.
(moving right) (moving down)

② When $x > p$,

If the x–values are increasing. $\Rightarrow$ the y–values are increasing.
(moving right) (moving up)

(2) $a < 0$

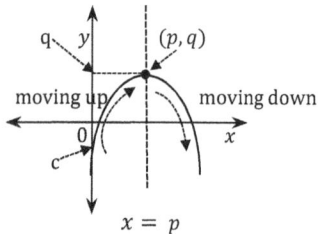

$x = p$

① When $x < p$,

If the x–values are increasing. $\Rightarrow$ the y–values are increasing.
(moving right) (moving up)

② When $x > p$,

If the x–values are increasing. $\Rightarrow$ the y–values are decreasing.
(moving right) (moving down)

5. Properties of $y = a(x - p)^2 + q$

(1) The signs of the x^2-coefficient

1) $a > 0$

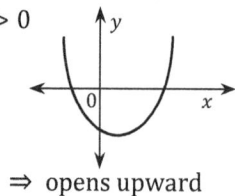

$\Rightarrow$ opens upward

2) $a < 0$

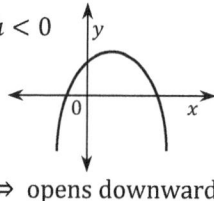

$\Rightarrow$ opens downward

$y = ax^2 + bx + c$

① Sign of a: $a > 0$ $a < 0$

② Sign of b: $ab > 0$ $ab < 0$ $b = 0$

③ Sign of c: $c > 0$ $c = 0$ $c < 0$

(2) The sides of the axis of Symmetry

$$y = a(x - p)^2 + q, \ a > 0 \ \Rightarrow \ \text{axis of symmetry} \quad x = p$$

1) $p > 0$

2) $p < 0$

3) $p = 0$

$\Rightarrow$ the y−axis is the axis of symmetry

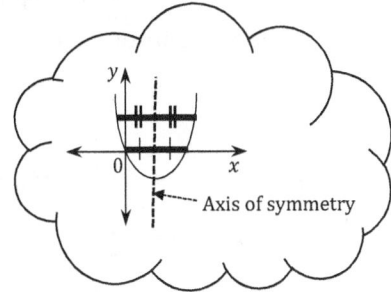

$x = p$

$x = p$

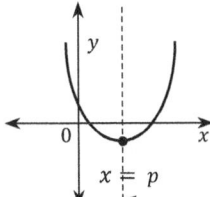

: on the right side of the y-axis

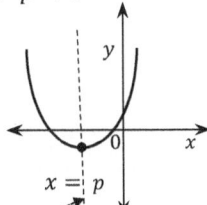

: on the left side of the y-axis

Axis of symmetry

Note : $y = ax^2 + bx + c \ \Rightarrow \ y = a\left(x + \dfrac{b}{2a}\right)^2 - \dfrac{b^2 - 4ac}{4a}$

$\Rightarrow \quad$ *Axis of symmetry* : $x = -\dfrac{b}{2a}$

① *if* $-\dfrac{b}{2a} > 0$ ($\dfrac{b}{2a} < 0$; $\dfrac{b}{a} < 0$; *a and b have different signs.*)

$\Rightarrow$ *The axis of symmetry is on the right side of the y -axis*

② *if* $-\dfrac{b}{2a} < 0$ ($\dfrac{b}{2a} > 0$; $\dfrac{b}{a} > 0$; *a and b have the same sign.*)

$\Rightarrow$ *The axis of symmetry is on the left side of the y -axis*

③ *if* $-\dfrac{b}{2a} = 0$

$\Rightarrow \ b = 0$

(3) The signs of the y-intercept

If $y = ax^2 + bx + c = a(x - p)^2 + q, \ a > 0 \ \Rightarrow$ the y-intercept is $c \ (= ap^2 + q)$.

① $c > 0$ $\Rightarrow$ the y−intercept is positive.	② $c < 0$ $\Rightarrow$ the y−intercept is negative.	③ $c = 0$ $\Rightarrow$ the y−intercept is 0.

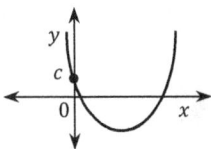

above the x-axis

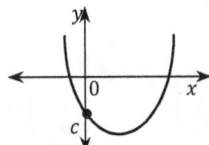

below the x-axis

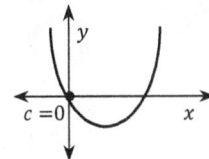

3-3 Solving Quadratic Functions

1. Equations of Quadratic Functions

(1) If the vertex (p, q) and one point are given,

$$\Rightarrow \quad y = a(x - p)^2 + q$$

Find a by substituting the point in the equation.

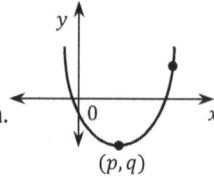

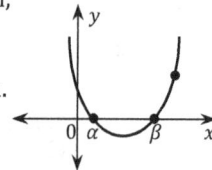

(2) If the axis of symmetry $x = p$ and two different points are given,

$$\Rightarrow \quad y = a(x - p)^2 + q$$

Find a and q

by substituting the two points in the equation.

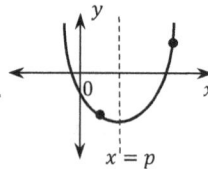

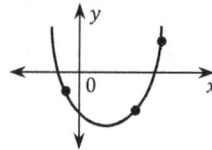

(3) If three different points are given,

$$\Rightarrow \quad y = ax^2 + bx + c$$

Find a, b and c

by substituting the points in the equation.

(4) If the x-intercepts α, β and one point are given,

$$\Rightarrow \quad y = a(x - \alpha)(x - \beta)$$

Find a by substituting the point in the equation.

2. Maximum or Minimum Values of Quadratic Functions

To get the maximum value or the minimum value of a quadratic function, transform the general form of the quadratic function $y = ax^2 + bx + c$ to the standard form of a quadratic function $y = a(x - p)^2 + q$.

(1) $y = a(x - p)^2 + q$

1) $a > 0$

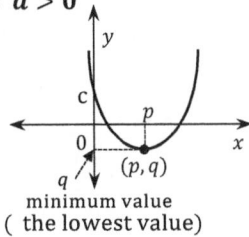

minimum value
(the lowest value)

When $x = p$,

$y = q$ is the minimum value of this quadratic function.

All the y-values of this function are greater than or equal to q ($f(x) \geq q$). Since the graph opens upward, there is no maximum value.

If $a > 0 \Rightarrow$ the minimum at the vertex, there is no maximum

If $a < 0 \Rightarrow$ the maximum at the vertex, there is no minimum

2) $a < 0$

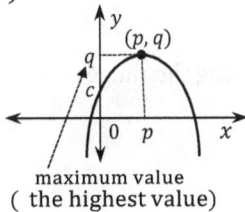

maximum value
(the highest value)

When $x = p$,

$y = q$ is the maximum value of this quadratic function.

All the y-values of this function are less than or equal to q ($f(x) \leq q$). Since the graph opens downward, there is no minimum value.

Note : $y = a(x - \alpha)(x - \beta)$

$\Rightarrow$ *This parabola has its maximum or minimum value at* $x = \dfrac{\alpha+\beta}{2}$

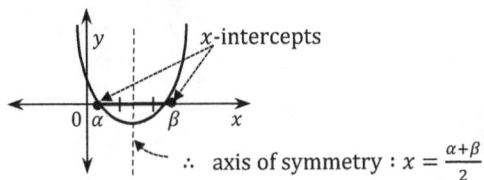

x-intercepts

$\therefore$ axis of symmetry : $x = \dfrac{\alpha+\beta}{2}$

(2) $y = ax^2 + bx + c$

First,
transform to the standard form !

$$y = ax^2 + bx + c = a\left(x + \frac{b}{2a}\right)^2 - \frac{b^2-4ac}{4a}$$

1) $a > 0 \;\Rightarrow\;$ minimum value : $-\dfrac{b^2-4ac}{4a}$, when $x = -\dfrac{b}{2a}$

2) $a < 0 \;\Rightarrow\;$ maximum value : $-\dfrac{b^2-4ac}{4a}$, when $x = -\dfrac{b}{2a}$

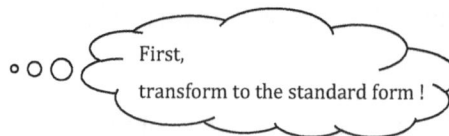

(3) When a Maximum Value or Minimum Value is given:

1) The minimum value m at $x = n$

$\Rightarrow y = a(x - n)^2 + m, \quad a > 0$

2) The maximum value m at $x = n$

$\Rightarrow y = a(x - n)^2 + m, \quad a < 0$

3. Steps for Solving Word Problems

(1) Determine the two variables x and y.

(2) Express the relationship between the variables x and y by considering the range of x-value.

(3) Simplify the function.

(4) Using the graph of the function, find the solution.

(5) Check the solution.

Exercises

#1 Identify the quadratic functions by marking O or ×.

(1) $y = \frac{1}{2}x^2 + 1$

(2) $y = 2x^2 - (3 + 2x^2)$

(3) $y = \frac{1}{x^2} + 1$

(4) $y = x^2 - (x + 1)^2$

(5) $y = x(x + 1)$

(6) $y = 2x^2 - x^2 + 1$

(7) $y = \frac{(x+1)^2}{3}$

(8) $2x^2 + 3x + 1$

(9) $y = 2$

(10) $y = 3x + 1$

#2 Find the following values of the quadratic function $f(x) = x^2 - 2x - 1$.

(1) $f(0)$

(2) $f(-1)$

(3) $f(2) + f(-2)$

(4) $f\left(-\frac{1}{2}\right)$

(5) $2f(1)$

#3 Find the value of $a + b$ for the quadratic equation $f(x) = -\frac{1}{2}x^2 + a$.

(1) $f(1) = -3$ and $f(-2) = b$

(2) $f(-1) = 1$ and $\frac{1}{2}f(0) = 2b$

(3) $\dfrac{f(1)+f(-1)}{2} = -\dfrac{1}{4}$ and $f(2) = -b$

#4 Find the vertex and the axis of symmetry for the following parabolas

(1) $y = 2x^2 - 4x$

(2) $y = x^2 - 2x - 3$

(3) $y = -x^2 - 2x + 2$

(4) $y = -\dfrac{1}{2}x^2 + 1$

(5) $y - 3 = 2(x - 2)^2$

#5   Identify the equations of the functions whose graphs are translated from the graph of $y = \frac{1}{2}x^2$ in the following ways

   (1) Translated -2 units along the x-axis

   (2) Translated 2 units along the y-axis

   (3) Translated 1 unit along the x-axis and -1 unit along the y-axis

   (4) Translated -3 units along the x-axis and -4 units along the y-axis

   (5) Translated m units along the x-axis and n units along the y-axis

#6   Find the value of a for which

   (1) The graph of $y = ax^2$ passes through one point $(2, -2)$.

(2) The graph of $y = \left(x - \frac{1}{2}\right)^2 + a$ passes through one point $(-1, 3)$.

(3) The graph of $y = \left(x + \frac{a}{2}\right)^2 + 3$ has the vertex $(-5, 3)$.

(4) The graph of $y = \left(x - \frac{2a}{3}\right)^2 - 2$ has been translated from the graph of $y = x^2 - 2$, -4 units along the x-axis.

(5) The graph of $y = 2(x + 3a - 1)^2 + 1$ has the y-axis as the axis of symmetry.

#7　Find the value of ab when the parabola $y = -ax^2 + b$

(1) Passes through $(-1, 2)$ and $(3, -2)$

(2) Passes through $(1, 2)$ and $(-2, -4)$

#8 Find the value of $a + b$ for the following graphs of quadratic functions

(1) $y = 2x^2 - x + 3$ passes through the two points $(1, a)$ and $(-2, -b)$.

(2) $y = -ax^2 + 2x - 1$ passes through the two points $(1, b)$ and $(-1, a)$.

(3) $y = -x^2 + 2ax + 3$ passes through the two points $(-1, 0)$ and $(2, b)$.

#9 For any constants m, n, the parabola $y = \frac{1}{2}(x - m)^2 + n$ is translated from $y = \frac{1}{2}x^2$. Give the conditions for m and n for the following parabola

(1)

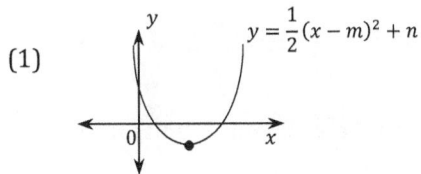
$y = \frac{1}{2}(x - m)^2 + n$

(2)

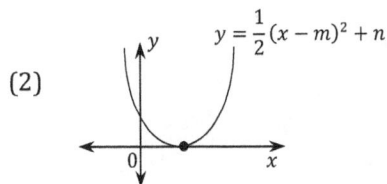
$y = \frac{1}{2}(x - m)^2 + n$

(3)

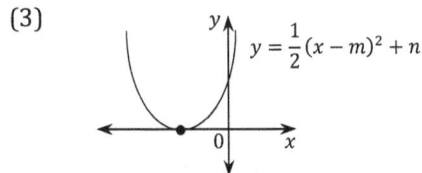
$y = \frac{1}{2}(x - m)^2 + n$

(4)

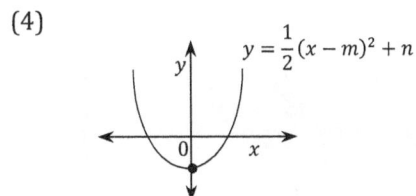
$y = \frac{1}{2}(x - m)^2 + n$

(5)

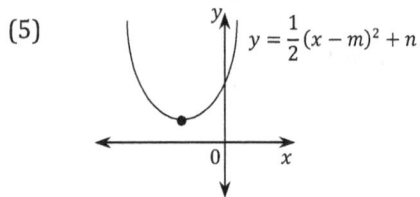

$$y = \frac{1}{2}(x - m)^2 + n$$

(6)

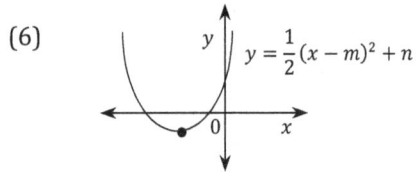

$$y = \frac{1}{2}(x - m)^2 + n$$

(7)

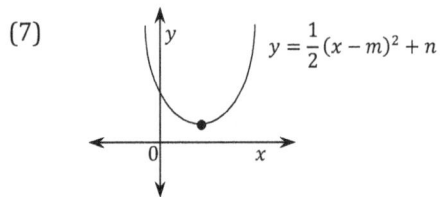

$$y = \frac{1}{2}(x - m)^2 + n$$

#10 Find an equation for the resulting quadratic function when

(1) The parabola $y = 2x^2 - 6x + 5$ is translated 2 units along the x-axis and -1 unit along the y-axis.

(2) The parabola $y = -3x^2 + 2x - 2$ is translated -2 units along the x-axis and 3 units along the y-axis.

(3) The parabola $y = -\frac{1}{2}(x + 2)^2 - 1$ is a symmetrical transformation along the x-axis.

(4) The parabola $y = \frac{1}{2}(x + 2)^2 + 1$ is a symmetrical transformation along the y-axis.

#11 Find the equation of the parabolas A and B on the graph.

Both are transformations of the parabola $y = \frac{1}{2}x^2$

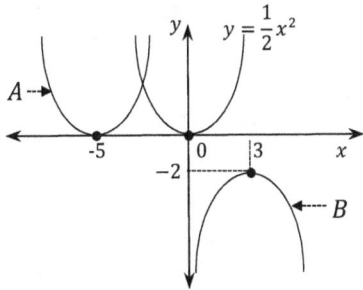

#12   State how the following parabolas have been translated from $y = -x^2 + 3x - 2$.

(1) $y = -x^2 + 4x + 3$

(2) $y = -x^2 + \frac{1}{2}x - 1$

(3) $y = -x^2 + 4x$

(4) $y = -x^2 + 3x + 2$

#13 Find the vertex, axis of symmetry, and intercepts for the following quadratic functions

(1) $y = 2x^2 + 3x + 1$

(2) $y = -x^2 + 2x + 3$

(3) $y = -3x^2 - 3x$

(4) $y = \frac{1}{2}x^2 - 4x + 6$

#14 Find the value of $a + p + q$ for the following quadratic functions

(1) $y = -3x^2 + 4x - a + 1 = a(x + p)^2 + q$

(2) $y = \frac{1}{2}x^2 - ax + 1 = \frac{1}{2}(x + 2)^2 + p + q$

(3) $y = ax^2 - 2x + 3 = -2(x + p)^2 - q$

#15 Find an equation of the quadratic function with following condition

(1) Vertex : $(1, 2)$ and passes through a point $(0, 3)$

(2) Axis of symmetry : $x = -1$ and passes through two points $(-3, -2)$, $(0, 4)$

(3) Vertex is on the x-axis, axis of symmetry : $x = -1$, and passes through a point $(-3, -4)$

(4) Passes through three points $(0, -3), (2, -1)$, and $(4, -6)$

(5) Passes through the origin, $(4, -3)$, and $(-2, 6)$

(6) Passes through $(-3, 0), (6, 0)$, and $(0, -6)$

(7) Passes through $(-4, 1), (-2, 0)$, and $(0, 3)$

#16 Find equations for the following parabolas

(1)

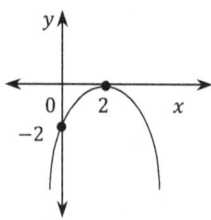

(2)

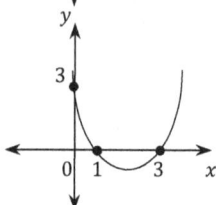

(3)

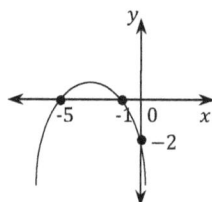

(4)

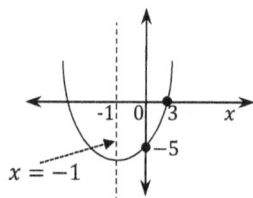

(5)

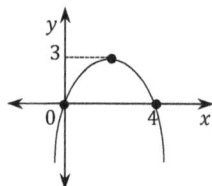

#17 Find the value of a for the following parabola

(1) The parabola $y = x^2 - ax + 2$ has $x = -2$ as its axis of symmetry.

(2) The parabola $y = -\frac{1}{2}x^2 + 4x - a + 1$ has its vertex on the x-axis.

(3) The distance between the two x-intercepts is 6 for a parabola $y = x^2 - 2x + a$.

#18 Find the minimum value or maximum values for the following quadratic functions

(1) $y = x^2 - 4x + 5$

(2) $y = -2x^2 + 4x + 1$

(3) $y = -3(x + 1)(x - 3)$

(4) $y = -x^2 + 4x - 4$

#19 Find the equation of the quadratic function with the following conditions

(1) The minimum value is 3 at $x = 1$ and passes through $(-2, 5)$.

(2) The maximum value is 4 at $x = -1$ and passes through $(1, -8)$.

#20 Find the value of $a + b$ for the following quadratic function with maximum or minimum values

(1) $y = -x^2 + 2ax + b$ has the maximum value 4 at $x = 1$.

(2) $y = 2x^2 - ax + b$ has the minimum value -3 at $x = -2$.

(3) $y = ax^2 + 2x + b$ has the maximum value 3 at $x = 2$.

(4) $y = ax^2 - bx + 2$ has the maximum value 3 at $x = -1$.

#21 One side of a rectangle is x inches. The perimeter and the area of a rectangle are 10 inches and y square inches, respectively. Find the maximum value of y.

#22 The sum of two numbers is 18. Find the maximum value of their product.

#23 The difference between two numbers is 10. Find the minimum value of their product.

#24 A ball is thrown upward from the top of a 5 foot table. After x seconds, the height of the ball from the ground is $y = -3x^2 + 12x + 5$. Find the maximum height from the ground the ball can reach.

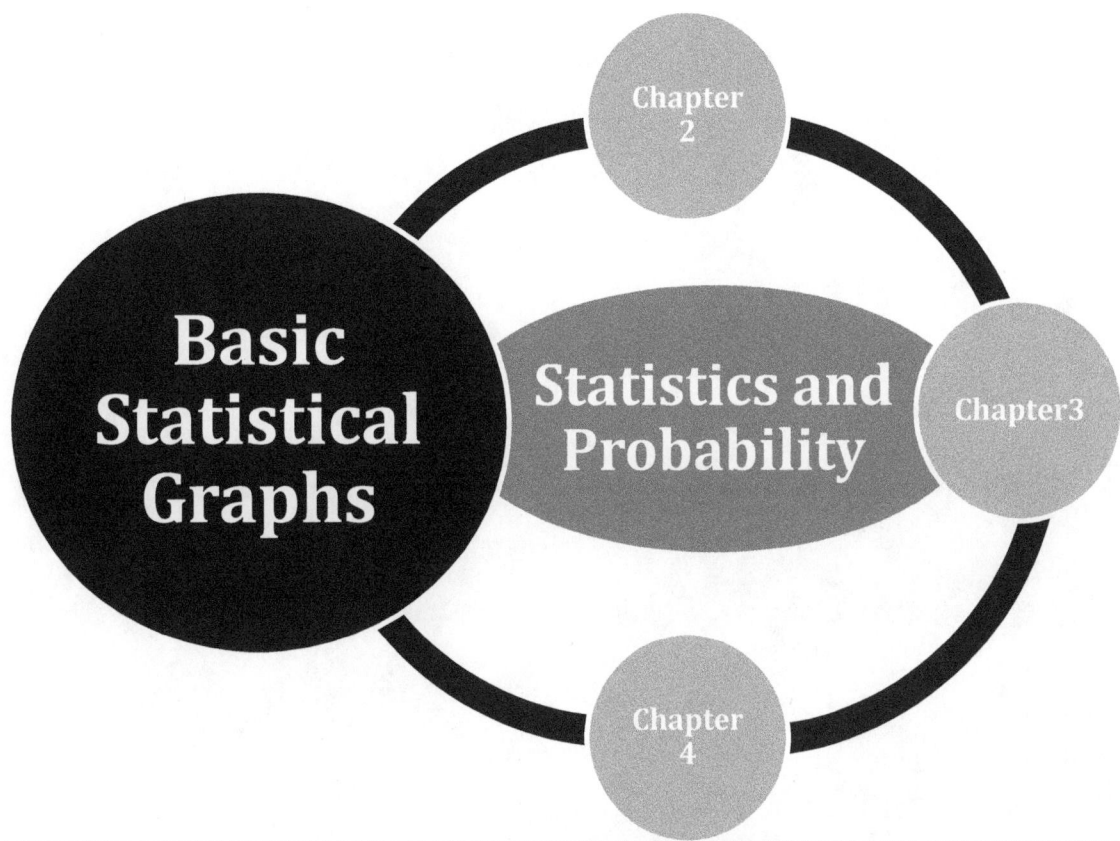

CHAPTER
1

Chapter 1 Basic Statistical Graphs

1-1 Categorical Data
1. Bar Graphs
2. Circle Graphs (Pie Charts)

1-2 Measurement Variables
1. Line Graphs
2. Scatter Plots

Chapter1. Basic Statistical Graphs

1-1 Categorical Data

Providing information in graphic forms makes a stronger visual impact. So, using graphs is often helpful in understanding data.

1. Bar Graphs

A *bar graph* displays frequencies as does a frequency histogram. The difference between a bar graph and a frequency histogram is that a bar graph places class limits on the horizontal axis.

Example   Consider the following data

66, 53, 41, 44, 63, 57, 61, 67, 72, 62

Class is a set of intervals (a range of values of a variable) into which the sample measurements may be grouped.

Class	Class limit	Measurements
40-50	41-50	41, 44
50-60	51-60	53, 57
60-70	61-70	61, 62, 63, 66, 67
70-80	71-80	72

A histogram represents numbers by area, not height.

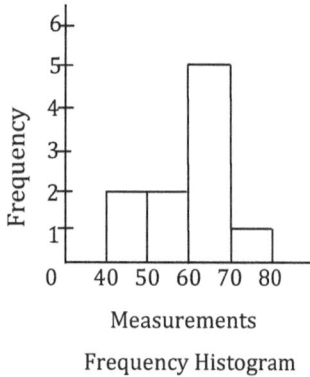

Frequency Histogram

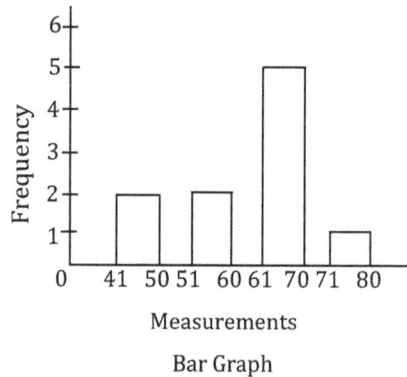

Bar Graph

We can place bars on a bar graph without any particular order because we don't place the class boundaries on the horizontal axis. Also, bar graphs can be used to represent two or more categorical measurements simultaneously. The height of each bar is proportional to the measurement value it represents.

2. Circle Graphs (Pie Charts)

Since a *circle graph* consists of parts, sectors, or sliced pieces in a circular pie, we can easily compare parts to each other and to the whole. By measuring the angle needed for each sector, we can draw the sectors correctly. The circle graph represents what percentage of the whole falls into each part.

Example Consider the following distribution of mid-term grades in a math class.

Grade	Number of students
A	3
B	8
C	5
D	1
F	1

The total number of students is $3 + 8 + 5 + 1 + 1 = 18$.

Now, find the angle needed for each sector.

Grade A : $\frac{3}{18} \times 360° = 60°$

Grade B : $\frac{8}{18} \times 360° = 160°$

Grade C : $\frac{5}{18} \times 360° = 100°$

Grade D : $\frac{1}{18} \times 360° = 20°$

Grade F : $\frac{1}{18} \times 360° = 20°$

The sum of the angle measure for all the sectors in a circle is 360°.

Using a protractor to measure angles, we can draw a precise circle graph.

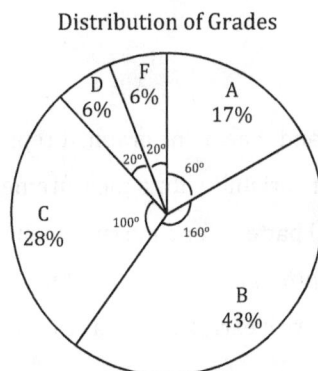

Distribution of Grades

We can more easily obtain information about the relative size of parts with a circle graph rather than with a table.

1-2 Measurement Variables

1. Line Graphs

These help to detect the pattern (trend)!

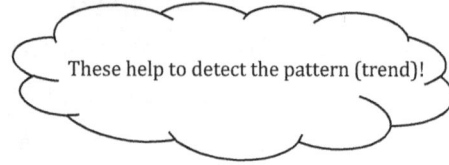

A *line graph* provides the same information as a histogram. Instead of using bars, we place a point at the correct height for each class interval and connect all the points with line segments. A steeper line segment represents that more change has occurred during that class interval. Observing a line graph, we can detect an overall pattern over a measurement variable so that we can make predictions.

Example Make a line graph for the following data

Number of books checked out from a library

42	38	47	52	60	45	36
Monday	Tuesday	Wednesday	Thursday	Friday	Saturday	Sunday

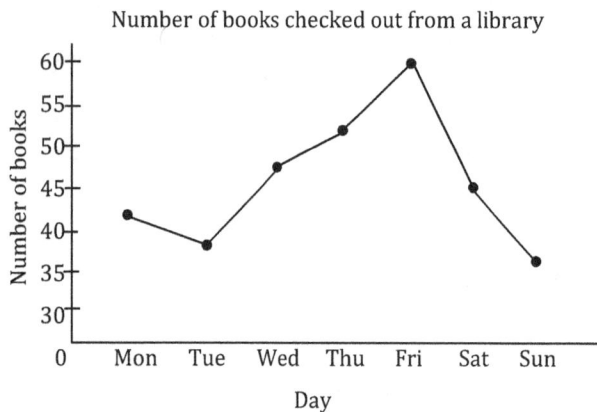

2. Scatter Plots

Even though a scatter plot can be more complicated to understand than a line graph, it represents more information about 3 important elements : outliers, the extent of variability for a measurement variable at each location of the other measurement variable, and an overall pattern for a measurement variable to increase as the other measurement variable increases. Thus, *scatter plots* are the most useful graphs for the relationship between two measurement variables. If variables are clearly related, a scatter plot can highlight the correlation between them.

A scatter plot is a plot of x_i and y_i as points (x_i, y_i) in a coordinate plane, where x_i and y_i are the values of measurement variables x and y for the same object.

For example, consider the following scatter plot displaying the relationship between the performance scores and daily practice minutes of every student in a class

Daily practice minutes vs. Performance scores

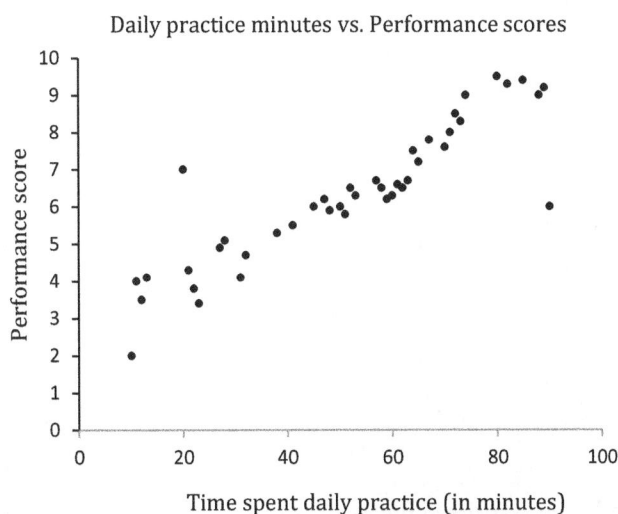

Each "•" on the plot represents a student.

The scatter plot shows an increasing trend toward higher scores with longer periods of practicing and also shows that there is still substantial variability in performance scores at each period of time spent in daily practice. Moreover, there are a few outliers. One student who only spent 20 minutes still earned 7 points in his/her performance, another student spent 90 minutes practicing and only earned 6 points in his/ her performance, the same score as several students who spent a significantly shorter period of time practicing.

Exercises

#1 Refer to the bar graph below to answer the following questions.

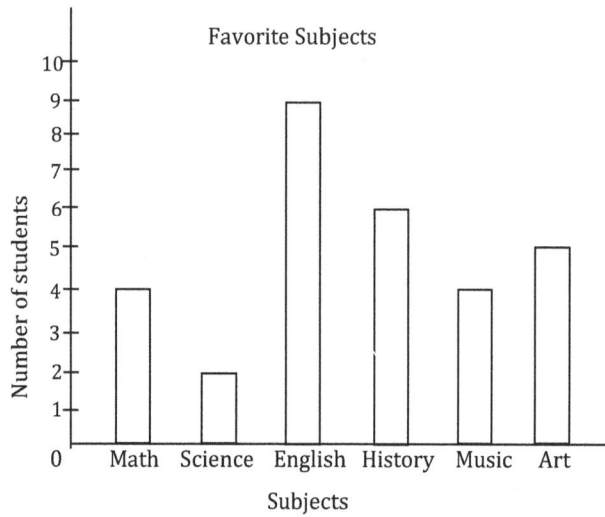

Favorite Subjects

(1) Which subjects are the most and least favorite subjects?

(2) Which two subjects are chosen by the same number of students?

(3) How many more students chose English than Math?

(4) What percentage of students chose history?

#2 Present the following information in a circle graph.

Item	Price
Shoes	$45
Pants	$30
Food	$20
Snack	$15
Bag	$10

#3 According to the circle graph, how much did Nichole spend on each item for the current month ?

Spending Expenses
Current Month
Total Income : $7,000

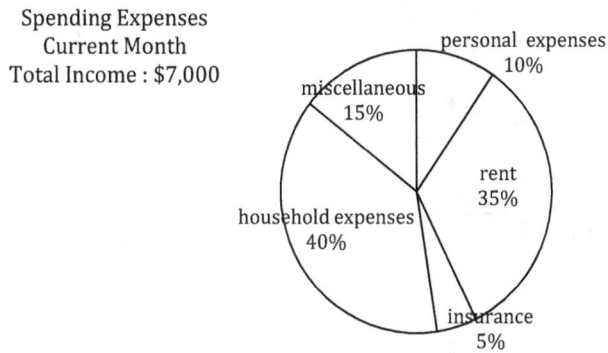

personal expenses
10%

miscellaneous
15%

rent
35%

household expenses
40%

insurance
5%

#4 Complete the following table using the categorical data listed in the table :

Category	Frequency	Relative Frequency	Central Angle(in degree)
A	50		
B	85		
C	65		
D	40		
Total	240	1.000	360

#5   Refer to the line graph below to answer the following questions.

Richard's time spent exercising

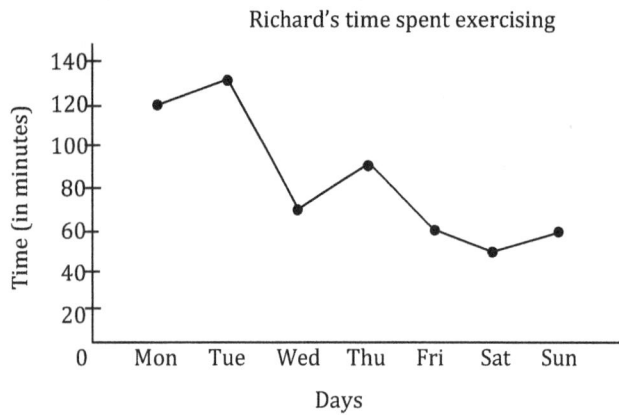

(1) On which day did Richard exercise most?

(2) Which days did Richard exercise for the same amount of time?

(3) Which two days show the greatest difference between Richard's time spent exercising?

Descriptive Statistics

Statistics and Probability

Chapter 1

Chapter 3

Chapter 4

CHAPTER 2

Chapter 2 Descriptive Statistics

Chapter 2. Descriptive Statistics

Statistics is the branch of scientific inquiry that provides methods for organizing and analyzing data. Studying statistics helps us to obtain an understanding from numbers. The study of statistics also includes the study of the development of techniques for collecting data.

2-1 Organizing Data

1. Stem Plots (Stem-and-Leaf Plots)

A *stem plot* is a convenient way to put a list of measurements into order while getting its shape. Stem plots provide a graphic description of summarizing data.

To make a stem plot for data,

(1) Separate each observation into a "stem" and "leaf". Generally, stems may have as many digits as needed, but each leaf should contain only one digit.

(2) List the stems in a vertical column from the smallest to the largest. Place a vertical line to the right of the stems, and add the leaves in the display row which corresponds to the observation's stem (from the smallest to the largest).

Using a stem plot, we can identify the center(median), which divides the data set into two sets of equal size. We can also identify the overall shape (symmetric or skewed in one direction) of the distribution. After observing the overall shape, we can check if outliers, or individual values that stand apart from the usual pattern, exist.

Example

 Consider the test scores of room 6 in a school.

90, 75, 82, 69, 77, 88, 88, 86, 90, 100, 84

List all the tens digits of each number as stems and the ones digits as leaves.

The resulting stem plot looks like this :

```
 6 | 9
 7 | 5  7
 8 | 2  4  6  8  8
 9 | 0  0
10 | 0
```

We find the center (median) of the distribution by counting the same amount of numbers from each end of the stem plot. That is, the center (median) is 86. The stem plot also shows that the distribution is symmetrical around the middle value.

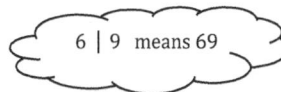

6 | 9 means 69

The stem plot is the simplest type of graphic description for a small number of observations in a data set. To create a graphic description of a large number of data sets, we use a relative frequency histogram which doesn't list all data values.

2. Relative Frequency Histogram

Frequency distribution provides a more compact summary of a data set than does a stem plot display. A graphical representation of a frequency distribution can be obtained by constructing a histogram. When frequencies are represented along the vertical axis, a histogram is called a frequency histogram.

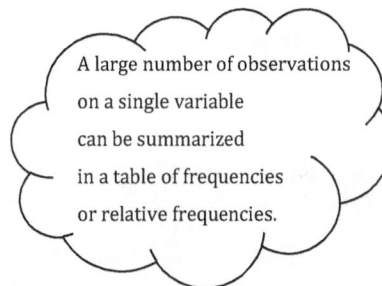

A large number of observations on a single variable can be summarized in a table of frequencies or relative frequencies.

To make a frequency histogram,

(1) Divide the *range* (the interval between the greatest and smallest measurement) into *classes* (equal subintervals). These classes must be specified precisely so that each measurement falls in to exactly one class.

(2) Count how many measurements fall into each class of the range.

Note : For the frequencies f_1, f_2, f_3, $\cdots\cdots$,

the quantities $\dfrac{f_1}{n}$, $\dfrac{f_2}{n}$, $\dfrac{f_3}{n}$, $\cdots\cdots$, where n is the total number of frequency,

are called relative frequencies.

Relative frequency is the proportion of observations falling in to the corresponding class interval.

(3) Draw a horizontal line to represent the measurement axis and a vertical axis to represent the frequency scale.

(4) Above each class interval, draw a bar with a height equal to the count for each class of the range. The heights of the bars are represented as proportions.

Example

A frequency distribution for test scores :

Class Interval (test scores)	Frequency (number of students)	Relative Frequency (number of students)
65-70	1	$\dfrac{1}{50}$
70-75	4	$\dfrac{4}{50}$
75-80	12	$\dfrac{12}{50} = \dfrac{6}{25}$
80-85	18	$\dfrac{18}{50} = \dfrac{9}{25}$
85-90	8	$\dfrac{8}{50} = \dfrac{4}{25}$
90-95	5	$\dfrac{5}{50} = \dfrac{1}{10}$
95-100	2	$\dfrac{2}{50} = \dfrac{1}{25}$
Total	50	

Frequency histogram for test scores :

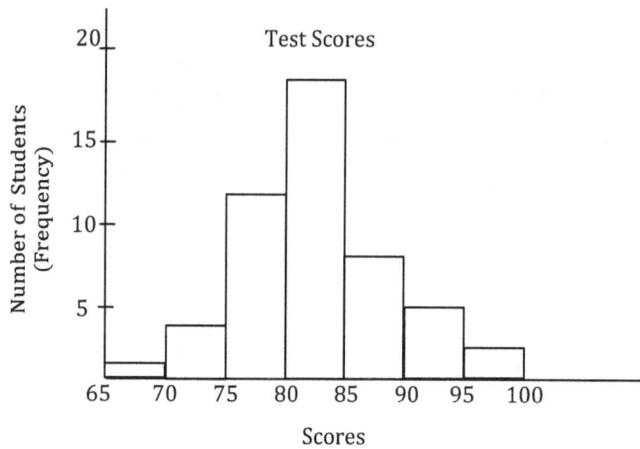

The distribution is centered in the 81 to 85 class. It is roughly symmetrical in shape.

The histogram shows that there are no large gaps or obvious outliers.

2-2 Measures of Central Tendency

Identifying the numerical measure of the center of a distribution is helpful when inspecting a stem plot or a histogram.

The most common numerical measure for describing the location of a data set is the arithmetic mean (or, also referred to simply as the mean).

1. The Mean

The most measure of central tendency (the tendency of the data to center about certain numerical values) for a quantitative data set is the mean of the data set.

The *mean* of a set of quantitative data is equal to the sum of all the sample measurements divided by the total number of measurements in the sample.

That is, for a given set of numbers $x_1, x_2, x_3, \cdots\cdots, x_n$, the mean is denoted by

$$\overline{x} = \frac{x_1 + x_2 + x_3 \ \cdots\cdots\ + x_n}{n}.$$

Example Consider the data set : 1, 2, 3, 4, 5, 6, 9, 10

Then, the mean is

$$\overline{x} = \frac{1+2+3+4+5+6+9+10}{8} = \frac{40}{8} = 5.$$

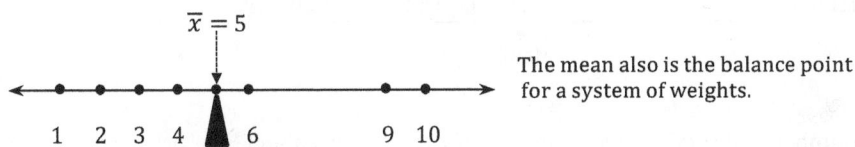

The mean also is the balance point for a system of weights.

Since the mean $\overline{x}$ represents the average value of the measurements in a sample data set, we can find the average, denoted by μ, of all values in the population.

The *population mean μ* is defined by

$$\mu = \frac{\text{Sum of } N \text{ population values}}{N}, \text{ for } N \text{ population.}$$

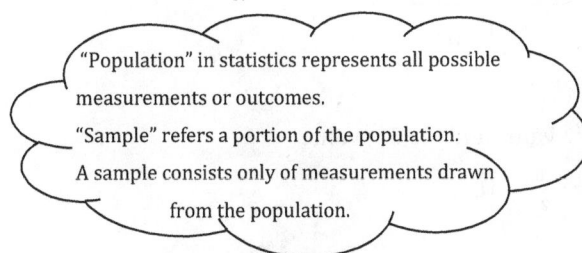

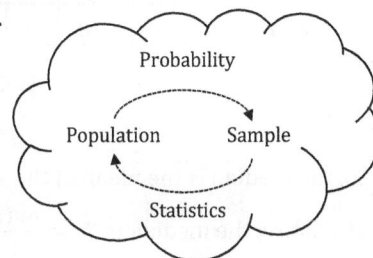

"Population" in statistics represents all possible measurements or outcomes.
"Sample" refers a portion of the population.
A sample consists only of measurements drawn from the population.

2. The Median

The mean as a measure of center is sensitive to the influence of the few extreme observations, also called outliers.

The median is a measure that is less sensitive to outlying values than the mean $\overline{x}$.

The *median* is the middle number of a distribution when the measurements in a data set are arranged in ascending order (from smallest value to largest value), including all repeated values.

If the number n of total measurements is odd, then the median M is the middle number in the ordered list. That is $M = \left(\frac{n+1}{2}\right)^{th}$ value in the ordered list.

If the number n of total measurements is even, then the median M is the mean of the two middle measurements in the ordered list. That is M is equal to the mean of the $\left(\frac{n}{2}\right)^{th}$ and $\left(\frac{n}{2}+1\right)^{th}$ ordered values.

Example

Consider the following data

15, 9, 7, 11, 8, 20, 11, 14, 9

To find the median, arrange all measurements in order from smallest to largest including repeated values. Then, the list of ordered values is

Order	1th	2nd	3rd	4th	5th	6th	7th	8th	9th
Measurement	7	8	9	9	(11)	11	14	15	20

Median

Since the number n of total measurements is $n = 9$ (odd number),

the median is the middle value, 11 (the $\left(\frac{n+1}{2}\right)^{th}$ value $= \left(\frac{9+1}{2}\right)^{th}$ value $= 5^{th}$ value).

If the largest measurement (20) had not appeared in this data, then $n = 8$ (even number).

Order	1th	2nd	3rd	4th	5th	6th	7th	8th
Measurement	7	8	9	9	11	11	14	15

Median $= \frac{9+11}{2} = 10$

So, the median is the mean of the 4th value and the 5th value.

Therefore, the median is $M = \frac{9+11}{2} = \frac{20}{2} = 10$.

3. The Mode

The *mode* is the measurement which occurs with the greatest frequency in a set of data.

The mode is of main value in describing large data set. The mode in a large data set will be in the class containing the largest relative frequency in a relative frequency histogram. Thus, the mode of a set of measurements is a useful measure of central tendency.

Example 1

Consider the following data

7, 2, 3, 3, 5, 3, 1, 2, 4, 2

Since 2 and 3 each occur three times (: more often than any other value), the modes are 2 and 3.

On the other hand, the data might consist of values, each of which occurs the same number of times.

Example 2

Consider the following data sets of A and B

A : 3, 7, 5, 1, 4

B : 6, 5, 6, 3, 5, 3

The two data sets have no mode.

Therefore, for a given set of data, it is possible to have more than one mode or no mode at all.

4. Comparing the Mean, Median, and Mode

The mean, median, and mode are all measures for the center of a data set. But, the mean and median focus on different aspects of the sample data. So, they are not equal generally.

For a population mean and a population median,

① Mode < Median < Mean

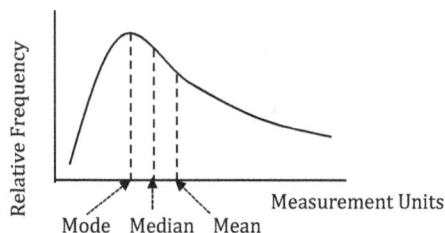

The data set is skewed to the right.
(positively skewed)
The higher values are more spread out than the lower values.

Note : A positive skew indicates that the tail on the right side is longer than the left side and the bulk of the values lie to the left of the mean.

② Mean = Median = Mode

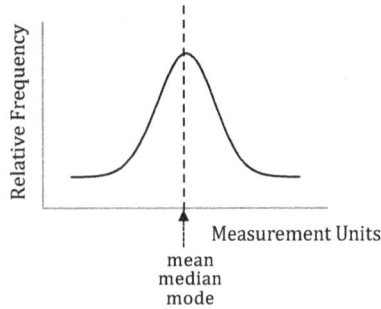

The data set is symmetrical

: Mirror image of the shape

   on the other side about the center line

: Bell curve data set

③ Mean < Median < Mode

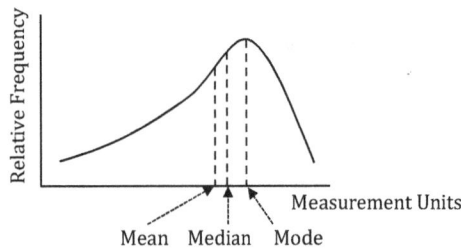

The data set is skewed to the left.

(negatively skewed)

: The lower values are more spread out

and the higher values tend to be

clumped.

Note : A negative skew indicates that the tail on the left side is longer than the right side and the bulk of the

values lie to the right of the mean.

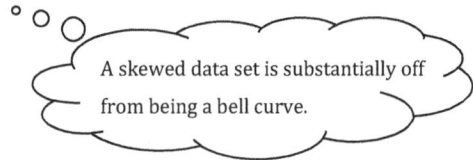

A skewed data set is substantially off from being a bell curve.

2-3 Measures of Variability (Spread)

1. The Range

Consider the following two histograms

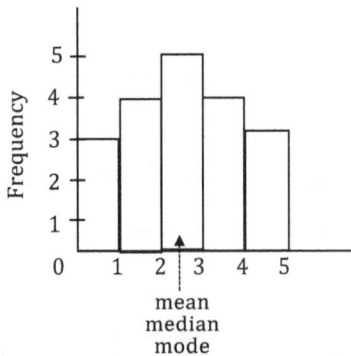

Figure 1

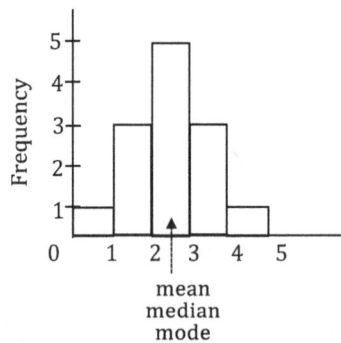

Figure 2

From the two histograms, we notice that both sets of data are symmetrical with equal means, medians, and modes. But Figure 1 and Figure 2 differ substantially in the variability or spread of the data set. Figure 1 shows an almost equal relative frequency, while most of the measurements in Figure 2 are clustered around their center. That is, the measurements in Figure 1 are more spread out or highly variable than the measurements in Figure 2 and the measurements in Figure 2 do not display much variability.

Therefore, we need to identify both a measure of relative variability as well as a measure of central tendency to describe these data sets.

The simplest measure of variability of a data set is its range.

The *range* of a data set is the arithmetic difference between the largest measurement and the smallest measurement.

Note : The ranges of the data sets in Figure 1 and Figure 2 are 5 – 3 = 2 and 5 – 1 = 4, respectively.
Thus, Figure 2 shows less variability than Figure 1.

Example
Consider the following data :

      72, 64, 66, 75, 59

Since the largest measurement of the data is 75 and the smallest measurement is 59,
the range of the data is 75−59=16.

2. Variance and Standard Deviation

Consider the following two sets of data A and B.

      A : 5, 8, 11, 25, 26, 27, 30, 45, 55, 68
      B : 5, 6, 7, 8, 10, 30, 43, 56, 67, 68

Since A and B have the same mean, 30, and range, 63, the two measures of central tendency are not distinguished between the two data sets. However, they have a distinctly different variability. The measurements in data set A cluster more closely around their center than do the measurements in data set B. Concerning an effective measure of variability, we consider the deviation and variance.

For any measurements $x_1, x_2, x_3, \cdots\cdots, x_n$,

$x_i - \overline{x}$ is called the *deviation* of the ith measurement from the mean, $\overline{x}$.

However,
$$(x_1 - \overline{x}) + (x_2 - \overline{x}) + \cdots\cdots + (x_n - \overline{x})$$
$$= (x_1 + x_2 + \cdots + x_n) - n \cdot \overline{x}$$
$$= (x_1 + x_2 + \cdots + x_n) - n \cdot \frac{(x_1 + x_2 + \cdots + x_n)}{n} = 0$$

That means, the average deviation from the mean is always 0.

Thus, we consider $(x_i - \overline{x})^2$ instead of $(x_i - \overline{x})$.

For any measurements $x_1, x_2, x_3, \cdots\cdots, x_n$,

the *variance* of a set of data, denoted by S^2, is defined by

$$S^2 = \frac{(x_1 - \overline{x})^2 + (x_2 - \overline{x})^2 + \cdots\cdots + (x_n - \overline{x})^2}{n}.$$

The standard deviation measures a set of data spreads out around its average. So, we can find how far away numbers on a list are from their average.

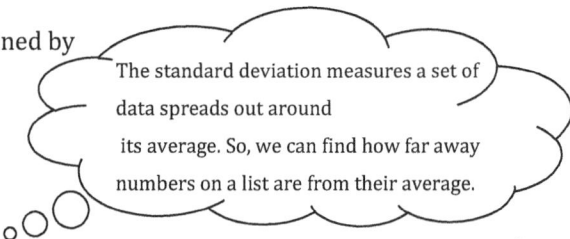

The *standard deviation* of a set of data, denoted by s, is the positive square root of the variance S^2.

That is, $s = \sqrt{S^2}$.

Variance and standard deviation are the most useful measures of variability.

Example

Consider the following data

2, 3, 4, 3, 5, 7

To obtain the variance and standard deviation, determine the mean, $\overline{x}$.

Since $\overline{x} = \frac{2+3+4+3+5+7}{6} = \frac{24}{6} = 4$,

the variance S^2 is

$$S^2 = \frac{(2-4)^2 + (3-4)^2 + (4-4)^2 + (3-4)^2 + (5-4)^2 + (7-4)^2}{6} = \frac{4+1+0+1+1+9}{6} = \frac{16}{6} = 2.67$$

and the standard deviation is $s = \sqrt{S^2} = \sqrt{2.67} \approx 1.63$

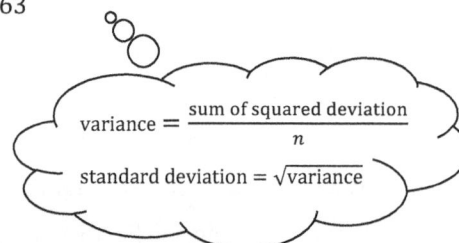

$$\text{variance} = \frac{\text{sum of squared deviation}}{n}$$

$$\text{standard deviation} = \sqrt{\text{variance}}$$

3. Quartiles

The variability or spread of a distribution can also be indicated by percentile.

Note : The median is the 50th percentile.

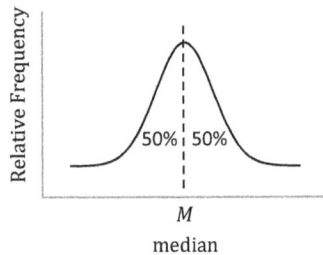

The most commonly used percentiles other than the median are the quartiles.

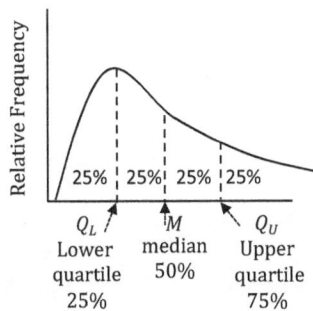

The quartiles divide a data set into four groups, each containing 25% of the measurements. The *lower quartile*, Q_L is the 25th percentile (the median of the smallest half of the data), the *middle quartile* is the 50th percentile(the median M), and the *upper quartile*, Q_U is the 75th percentile (the median of the largest half of the data).

Example

Consider the following data

      1, 2, 2, 3, 4, 5, 5, 6, 7

The data set is arranged in order and the number n of measurements is $n = 9$ (odd).

So, the median M is the middle value $\left(\frac{n+1}{2} = \frac{9+1}{2} = 5^{th} \right)$, 4.

Since the median M divides the data set into two equal halves, the lower quartile is the median of the lower half : 1, 2, 2, 3 and the upper quartile is the median of the upper half : 5, 5, 6, 7.

Since each half contains $n = 4$ (even), of measurements, the median is the mean of two middle values.

That is, the lower quartile is $\frac{2+2}{2} = 2$ and the upper quartile is $\frac{5+6}{2} = 5.5$

Quartiles are found
in ascending order.

4. Boxplots (Box-and-Whisker Plots)

Boxplots provide a method for detecting outliers (measurements that are further away from the rest of the data) based on the distance between the lower quartile, Q_L and the upper quartile, Q_U.
Interquartile range (IR) is defined by $IR = Q_U - Q_L$.

Boxplots show the center (median), variability (spread), the extent of skewness from symmetry, and the detection of outliers.

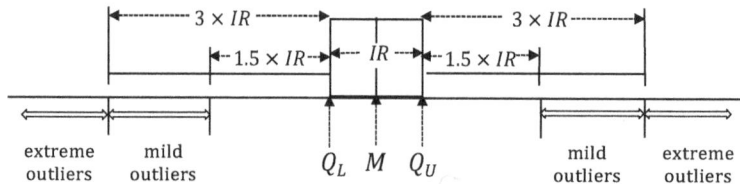

Example

Consider the following data

0, 1, 5, 5, 7, 8, 9, 10, 24, 33

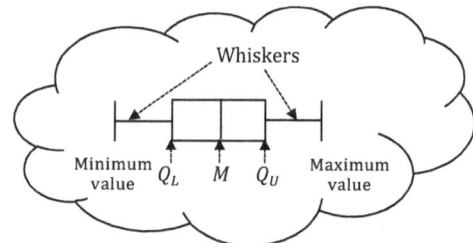

Since the sample size is $n = 10$ (even), the median M is the mean of $\left(\frac{n}{2}\right)^{th}$ and $\left(\frac{n}{2} + 1\right)^{th}$ values.

That is, $M = \frac{7+8}{2} = 7.5$ The lower quartile Q_L is 5 and the upper quartile Q_U is 10.

Order	1th	2nd	3rd	4th	5th	6th	7th	8th	9th	10th
Measurement	1	1	⑤	5	7	8	9	⑩	24	33

Q_L M Q_U

So, $IR = Q_U - Q_L = 10 - 5 = 5$.

Since $1.5 \times IR = 7.5$ and $3 \times IR = 15$, the boxplot is

$Q_L = 5$ $M = 7.5$ $Q_U = 10$ 17.5   25

From the boxplot, we see

① The median is in the middle of the box.

② The middle 50% (from Q_L to Q_U) of the data set shows that the data forms a symmetrical distribution.

③ There is a mild outlier (indicated by open circle), 24, and an extreme outlier (indicated by closed circle), 33.

5. Line Plots

If a set of data has distinct measurements, we use a line plot with an $\times$ for each measurement whenever it appears in the set of data. Line plots display the frequencies in the data set, as does a frequency histogram.

Using a line plot, we can also obtain the mode, median, range, and outliers.

Example

Consider the following frequency distribution of measurements.

Measurement	Frequency	Relative Frequency
1	4	$\frac{4}{20} = \frac{1}{5} = 20\%$
2	2	$\frac{2}{20} = \frac{1}{10} = 10\%$
3	6	$\frac{6}{20} = \frac{3}{10} = 30\%$
4	4	$\frac{4}{20} = \frac{1}{5} = 20\%$
5	2	$\frac{2}{20} = \frac{1}{10} = 10\%$
6	2	$\frac{2}{20} = \frac{1}{10} = 10\%$

Frequency distribution

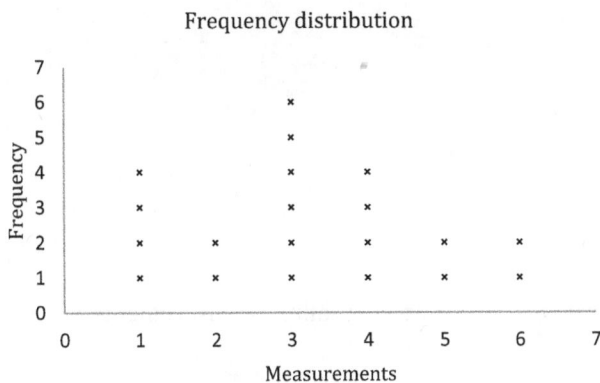

From the line plot, we observe that the mode is 3, the median is 3, and the range is 6−1=5.

The mean is $\frac{1\cdot4+2\cdot2+3\cdot6+4\cdot4+5\cdot2+6\cdot2}{20} = \frac{64}{20} = 3.2$

Exercises

#1 Make a stem-and-leaf plot for the following data.

43, 100, 50, 64, 73, 79, 81, 66, 55, 61, 101, 52, 55, 48, 64, 113, 77, 80, 81, 95, 53

#2 The following are the hourly wages in dollars of 20 workers. Arrange the data given below in a frequency table.

9.80 9.60 10.15 9.80 10.60 12.20 8.85 11.50 9.60 10.20

10.15 8.85 9.80 10.20 10.15 9.80 11.50 9.80 9.80 9.60

#3 The following distribution shows the number of days in a year each student in a class of 50 visited a doctor. Draw a relative frequency histogram.

Number of days	Number of students
1-5	4
6-10	6
11-15	11
16-20	13
21-25	9
26-30	7

#4 Calculate the mean, median, mode, and range for the following sample measurements

(1) 3, 6, 7, 5, 4, 3, 2

(2) 9, 3, 5, 5, 2, 20, 4, 6

(3) 23.5, 31.2, 18.4, 35.4, 25

(4) -4, -5, -7, -3, -4, -7, -2, -6

#5 Find the deviations, variance, and standard deviation for the following data

25, 30, 36, 38, 43, 56

#6   Refer to the data set below to answer the following questions.

   27, 26, 27, 38, 23, 27, 39, 42, 38, 63, 34

   (1) Determine the lower quartile, Q_L and upper quartile, Q_U .

   (2) Calculate the value of the interquartile range (IR).

   (3) Draw a box plot.

#7 Complete the following table and draw a line plot for the frequency distribution.

Measurement	Frequency	Relative Frequency (%)
36.4	3	
42.5	6	
48.1	4	
53.2	6	
55.8	8	
64.7	9	
66.3	6	
72.9	3	
total	45	1 = 100%

#8 Refer to the data set below to answer the following questions.

Age of teachers in a certain school.

43 40 55 47 37 36 52 40 28 26 42 40 36 45 39

(1) Draw a stem-and-leaf plot for the data.

(2) Complete the frequency table.

Age	Frequency	Relative Frequency (%)
26-30		
31-35		
36-40		
41-45		
46-50		
51-55		

(3) Draw a relative frequency histogram.

(4) Find the mean, median, mode, and range.

(5) Draw a line plot.

(6) Draw a box plot.

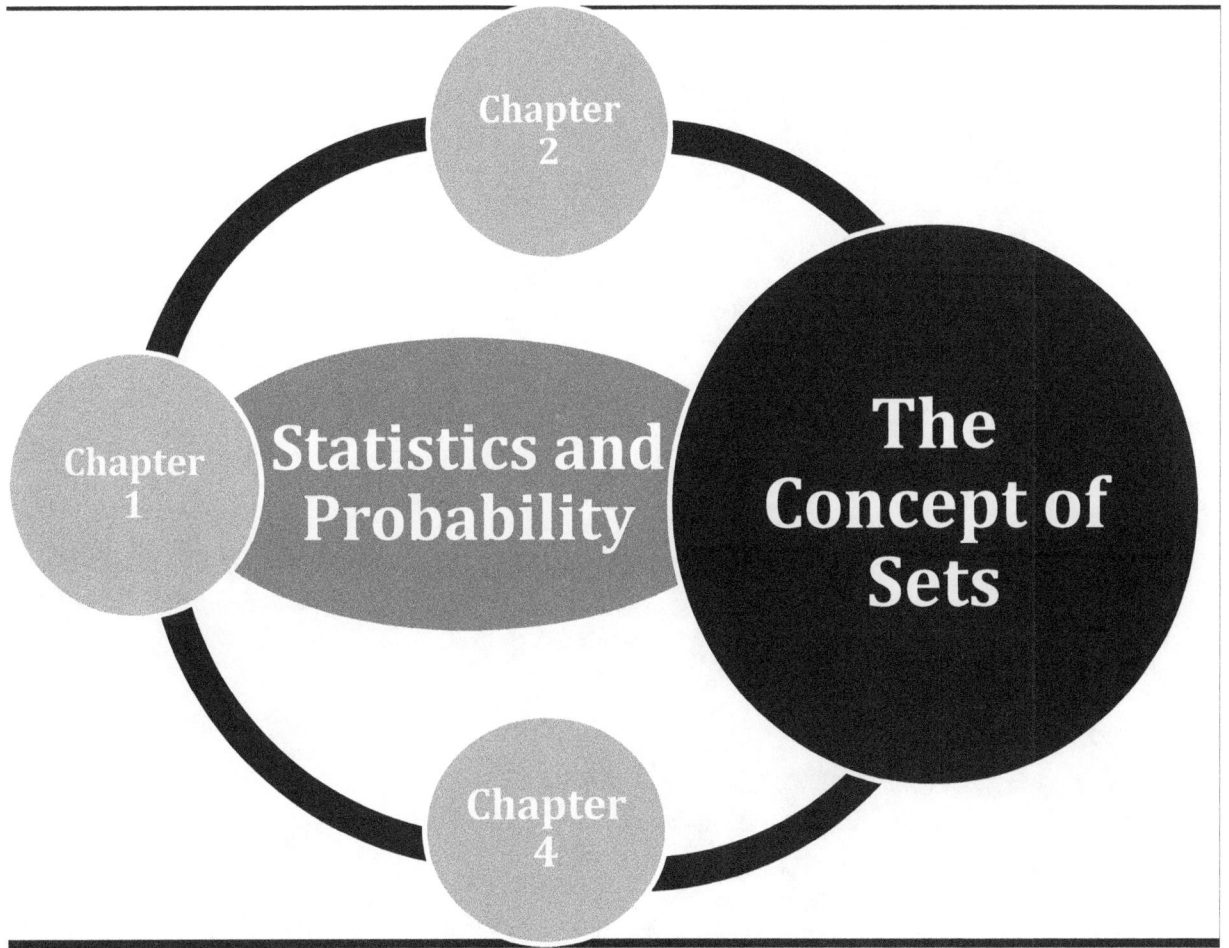

Chapter
2

Chapter
1

Statistics and
Probability

The
Concept of
Sets

Chapter
4

Chapter 3   The Concept of Sets

3-1 Sets

1. **Definition**

    (1) Finite Sets

    (2) Infinite Sets

    (3) Empty Set

2. **Symbols**

3. **Set Notation**

3-2   **Subsets**

1. **Definition**

    (1) Identical

    (2) Subset

    (3) Superset

    (4) Proper Subset

2. **Properties of Subsets**

3. **Numbers of Subsets**

3-3 **Operations on Sets**

1. **Unions and Intersections**

    (1) Definition

    (2) The Property of Unions and Intersections

    (3) The Number of Elements

2. **Complements**

    (1) Definition

    (2) Properties of Complements

    (3) The Number of Elements

Chapter 3. The Concept of Sets

3-1 Sets

1. Definition

A *set* is any collection into a whole of definite, distinguishable objects, called *element*s.

For example,   ① The set of all books in a library.

② The set of all students in a classroom.

③ The set of letters *a, b*, and *c.*

④ The set of rules in a group.

⑤ The set of all positive integers whose square is 2.

⑥ The set of all numbers greater than 3.

(1) Finite Sets

A finite set is a set which contains only a finite number of elements.

Examples ① to ⑤ above are all finite sets.

(2) Infinite Sets

An infinite set is a set which contains an infinite number of elements

Example ⑥ above is an infinite set.

(3) Empty Set

An empty set is a set which has no elements.

Example ⑤ above is an empty set.

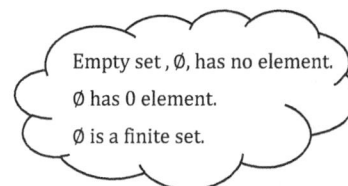

> Empty set , ∅, has no element.
>
> ∅ has 0 element.
>
> ∅ is a finite set.

2. Symbols

Sets are designated by the following symbols.

(1) { } ： The symbol { } represents the elements in a set.

The set example ③ above is $\{a, b, c\}$, and the set example ⑥ above is $\{4, 5, 6, \cdots \cdots \}$.

(2) ∅ ： An empty set is denoted by the symbol ∅.

The set example ⑤ above is ∅.

(3) $\in$, $\notin$: The symbols $\in$, $\notin$ designate whether an element belongs (or doesn't belong) to a set.

    ① If a is an element of the set A, then we write $a \in A$ (a belongs to A.).

    ② If a is not an element of the set A, then we write $a \notin A$ (a does not belong to A.).

    For example, $A = \{1, 2, 3\} \Rightarrow 3 \in A,\ 4 \notin A$

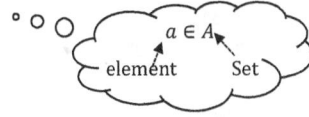

(4) $n(A)$: The number of elements in a set A is denoted by $n(A)$.

    For example, ① If A is the set of all prime numbers less than 10, then

$$n(A) = n(\{2, 3, 5, 7\}) = 4$$

    ② If $A = \emptyset$, then $n(A) = n(\emptyset) = 0$

3. Set Notation

(1) A set $A = \{a, b, c\}$ is the same as $\{b, a, c\}$ or $\{c, b, a\}$, etc.

Note : *① $\{a, a, b\}$ is not a proper notation of a set, because all the elements in a set are distinct.*
It should be replaced by $\{a, b\}$.

    ② $\{a, b, c, \cdots\cdots\}$ is used when a set includes many elements and there is a fixed rule between the
elements. For example, a set of even numbers is $\{2, 4, 6, \cdots\cdots\}$.

(2) The set builder notation : $\{x \in A \mid P(x)\}$

$\{x \in A \mid P(x)\}$ is the set of all x in A such that $P(x)$ is true.

As a rule, to every set A and to every statement $P(x)$ about $x \in A$, there is a set $\{x \in A \mid P(x)\}$ whose elements are precisely those elements x of A for which the statement $P(x)$ is true.

Example

Let A be the set of all students in a class.

The statement "x is a boy." is true for some elements x of A and false for others.

To specify the set of all the boys in the class, we use the notation $\{x \in A \mid x$ is a boy. $\}$.

Similarly, $\{x \in A \mid x$ is not a boy. $\}$ specifies the set of all the girls in the class.

(3) Venn Diagram

Using a diagram called the Venn Diagram, the relationship between sets are visualized and illustrated.

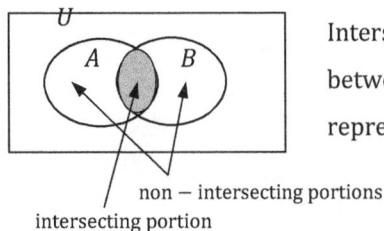

Intersecting portion of two circles A and B represents similarities between two sets, while non-intersecting portions of the circles represent differences between two sets.

non − intersecting portions

intersecting portion

3-2   Subsets

1. Definition

(1) Identical

Two sets are called *identical*

if every element of each set is an element of the other set. Identical sets are denoted by $A = B$.

Note : If two sets A an d B are equal or identical, then A = B (Both sets contain the same element).

(2) Subset

If all elements of a set A are elements of a set B,

then set A is called a *subset* of a set B and denoted by $A \subseteq B$ or $B \supseteq A$.

(3) Superset

If a set A is a subset of a set B, then B is called a *superset* of A.

(4) Proper Subset

If $A \subseteq B$ and $A \neq B$,

then we write $A \subset B$ or $B \supset A$ and

A is called a *proper subset* of B or B is called a *proper superset* of A.

If A is not a subset of B, then we denote $A \nsubseteq B$.

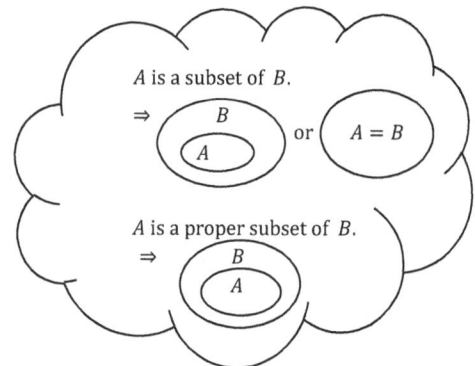

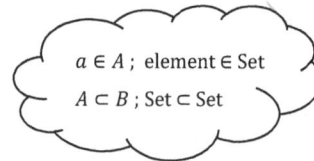

A is a subset of B.
$\Rightarrow$ B A or $A = B$

A is a proper subset of B.
$\Rightarrow$ B A

2. Properties of Subsets

(1) The empty set $\emptyset$ is a subset of every set.

$\emptyset \subset A, \ \emptyset \subset \emptyset$

(2) Every set is a subset and a superset of itself.

$A \subseteq A$

$a \in A$; element $\in$ Set
$A \subset B$; Set $\subset$ Set

(3) If $A \subseteq B$ and $B \subseteq C$, then $A \subseteq C$.

(4) If $A \subset B$ and $B \subset A$, then $A = B$.

3. Numbers of Subsets

For any set A which has $n(A) = m$,

A proper subset does not include itself.

(1) The number of subsets is $2^m = \underbrace{2 \times 2 \times \cdots \cdots \times 2 \times 2}_{m \text{ times}}$

(2) The number of proper subsets is $2^m - 1$

$n(\text{set})$
$n(\emptyset) = 0$
But $n(\{\emptyset\}) = 1$ and $n(\{0\}) = 1$

set with an element $\emptyset$

set with an element 0

Example $A = \{a, b, c\}$

The subsets of A are $\emptyset$, $\{a\}$, $\{b\}$, $\{c\}$, $\{a, b\}$, $\{a, c\}$, $\{b, c\}$, $\{a, b, c\}$.

　　　　　　　　　　0-element　1-element　　2-elements　　3-elements

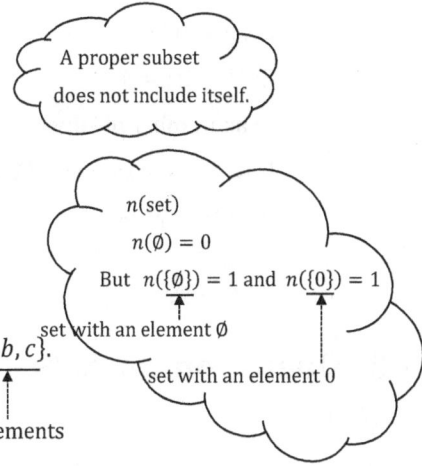

So, the set A has 8 subsets.

Since $n(A) = 3$, the number of subsets is $2^3 = 8$.

Since the proper subset does not include itself, the number of proper subsets is $2^3 - 1 = 7$.

Note : ① $A = \emptyset \Rightarrow n(A) = 0$ *and* $n(A) = 0 \Rightarrow A = \emptyset$

② $A \subset B \Rightarrow n(A) \leq n(B)$ *but*, $n(A) \leq n(B) \nRightarrow A \subset B$

$(\because \textit{for example}, \ A = \{a, b\}, \ B = \{b, c, d\}$

$\Rightarrow n(A) = 2 < 3 = n(B) \ \textit{but} \ A \not\subset B \ )$

③ $A = B \Rightarrow n(A) = n(B)$ *but*, $n(A) = n(B) \nRightarrow A = B$

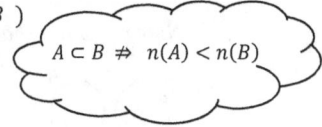

$A \subset B \nRightarrow n(A) < n(B)$

(3) The number of subsets which do include (or do not include) certain elements is 2^{m-p}, where $p = n(\{\text{cetain elements}\})$

Example $A = \{d, e, f\}$

① The number of subsets which include d :

To find the subsets which include the element "d", first consider the subsets of $\{e, f\}$ which exclude the element "d". Then, put the element "d" back in the subsets.

So, the number of subsets which include the element "d" is the number of subsets of $\{e, f\}$.

$2^{3-1} = 2^2 = 4$

② The number of subsets which do not include d :

Find the subsets excluding the element "d".

So, it will be the same as the number of subsets of $\{e, f\}$.

$2^{3-1} = 2^2 = 4$

③ The number of subsets which include d but not e:

It will be the same as the number of subsets of $\{f\}$.

$2^{3-1-1} = 2^1 = 2$

> $A = \{a, b, c, d, e\}$
> The number of subsets
> which include a, b but not c is
> $2^{5-2-1} = 2^2 = 4$.

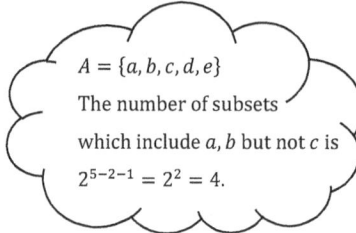

3-3 Operations on Sets

1. Unions and Intersections

(1) Definition

1) **Union** : The *union* of any two sets A and B is the set of all elements x such that x belongs to at least one of the two sets A and B . The union is denoted by $A \cup B$.

$A \cup B = \{ x \in A \cup B \mid x \in A \text{ or } x \in B \}$

Note : $A \cup B$ is shaded.

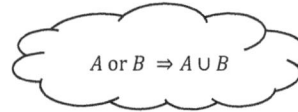

Figure 1   Figure 2   Figure 3

> A or $B \Rightarrow A \cup B$

2) **Intersection** : The *intersection* of any two sets A and B is the set of all elements x which belong to both A and B . The intersection is denoted by $A \cap B$.

$A \cap B = \{ x \in A \cap B \mid x \in A \text{ and } x \in B \}$

Note : $A \cap B$ is shaded.

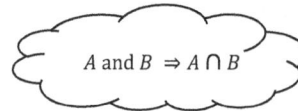

Figure 1   Figure 2   Figure 3

> A and $B \Rightarrow A \cap B$

(2) The Property of Unions and Intersections

For any sets A and B,

1) $A \cup \emptyset = A, \quad A \cap \emptyset = \emptyset$

2) $A \cup A = A, \quad A \cap A = A$

3) $A \cup B = B \cup A, \quad A \cap B = B \cap A$

4) $A \subset B \Rightarrow A \cup B = B, A \cap B = A$

5) $(A \cap B) \subset A \subset A \cup B, \quad (A \cap B) \subset B \subset A \cup B$

(3) The Number of Elements

For any finite sets A and B,

1) $n(A \cup B) = n(A) + n(B) - n(A \cap B)$

2) $n(A \cap B) = n(A) + n(B) - n(A \cup B)$

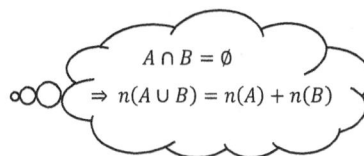

$A \cap B = \emptyset$

$\Rightarrow n(A \cup B) = n(A) + n(B)$

Example

$A = \{a, b, c\}, \quad B = \{b, c, d, e\}$

$\Rightarrow A \cup B = \{a, b, c, d, e\}$ and $A \cap B = \{b, c\}$

Using a Venn Diagram,

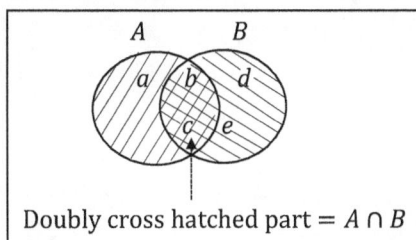

Doubly cross hatched part $= A \cap B$

$A \cap B$ belongs to A and also B. That means $n(A \cap B)$ is doubly added to $n(A) + n(B)$.

So, $n(A \cap B)$ should be subtracted one time from $n(A) + n(B)$.

$n(A \cup B) = n(A) + n(B) - n(A \cap B)$.

Note : For any finite sets $A, B,$ and C,
$$n(A \cup B \cup C) = n(A) + n(B) + n(C) - n(A \cap B) - n(B \cap C) - n(A \cap C) + n(A \cap B \cap C).$$

2. Complements

The operation for complementation is similar to the operation of subtraction.

(1) Definition

1) A^C : For any subset A of the universal set U, the set of all the elements which do not belong to A is denoted by A^C. $A^C = \{x \mid A \subset U,\ x \in U,\ x \notin A \}$

   Note : For a fixed set U which may be regarded as a universal set, $U - A = A^C$

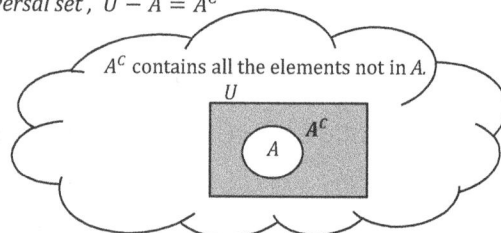

   A^C contains all the elements not in A.

2) $A - B$: The relative complement of a set B in a set A is the set $A - B$ and it is denoted by $A - B$. $A - B = \{x \mid x \in A \text{ and } x \notin B \}$

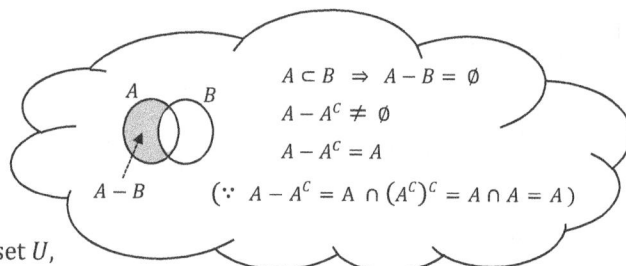

(2) Properties of Complements

$A \subset B \Rightarrow A - B = \emptyset$

$A - A^C \neq \emptyset$

$A - A^C = A$

$A - B$ $(\because A - A^C = A \cap (A^C)^C = A \cap A = A)$

For any subsets A and B of the universal set U,

1) ① $(A^C)^C = A$

   ② $\emptyset^C = U,\ U^C = \emptyset$

   ③ $A \cup A^C = U,\ A \cap A^C = \emptyset$

   ④ $A \subseteq B$ if and only if $B^C \subseteq A^C$

   Note : De Morgan's theorem

For any two sets A and B,	① $(A \cup B)^C = A^C \cap B^C$
	② $(A \cap B)^C = A^C \cup B^C$

2) ① $U - A = A^C, \quad U - \emptyset = U$

② $A - A = \emptyset, \quad A - \emptyset = A$

③ $A - B = A \cap B^C = A - (A \cap B) = A \cup B - B$

④ $A - B = \emptyset \Rightarrow A \subseteq B$

⑤ $A \cap B = \emptyset \Rightarrow A - B = A, \quad B - A = B$

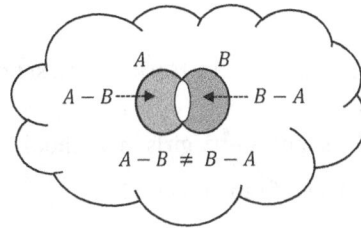

A B

$A - B$ ---→   ←--- $B - A$

$A - B \neq B - A$

(3) The Number of Elements

For any subsets A and B of the universal set U,

① $n(A^C) = n(U) - n(A)$

② $n(A - B) = n(A) - n(A \cap B) = n(A \cup B) - n(B)$

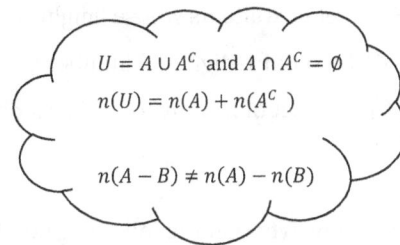

$U = A \cup A^C$ and $A \cap A^C = \emptyset$

$n(U) = n(A) + n(A^C)$

$n(A - B) \neq n(A) - n(B)$

Note : $R = \{x \mid x \text{ is a real number}\}$

$Q = \{x \mid x \text{ is a rational number}\}$

$Z = \{x \mid x \text{ is an integer}\}$

$N = \{x \mid x \text{ is a natural number}\}$

$\Rightarrow \quad N \subset Z \subset Q \subset R$

Exercises

#1 Identify all the sets. Mark o for a set or × for a non-set.

    (1) A set of pretty girls in a school.

    (2) A set of red apples.

    (3) A set of natural numbers.

    (4) A set of small numbers.

    (5) A set of famous singers.

    (6) A set of even numbers.

    (7) A set of people who like math.

    (8) A set of students whose heights are less than 5 feet in a class.

    (9) The set of 1-digit odd numbers.

    (10) The set of healthy foods in a store.

#2 Determine whether the following notations are true or false.

    (1) $\{1, 2, 3\} = \{3, 1, 2\}$

    (2) $\{1, 2, 3, 4, 5\} = \{x \mid x \text{ is a natural number less than } 6.\}$

    (3) $\{x \mid x \text{ is a factor of } 6.\} = \{1, 2, 3, 6\}$

    (4) $\{x \mid x \text{ is a natural number less than } 1.\} = \{0\}$

    (5) $\{0, 4, 8, 12, 16 \cdots\} = \{x \mid x \text{ is a multiple of } 4.\}$

    (6) $\{1, 3, 5, 7, 9\} = \{x \mid x \text{ is an odd number less than } 10.\}$

    (7) $\{x \mid 1 \leq x \leq 3, \ x \text{ is an integer.}\} = \{1, 2, 3\}$

    (8) $\{1, 2, 3, 4\} = \{x \mid x \text{ is a prime number less than } 5.\}$

#3 State if the following sets are finite or infinite sets.

    (1) $\{x \mid x \text{ is a factor of } 20.\}$

    (2) $\{x \mid x \text{ is a multiple of } 2.\}$

    (3) $\{x \mid x \text{ is an even number.}\}$

    (4) $\{x \mid 1 \leq x \leq 3, \ x \text{ is an odd number.}\}$

    (5) $\{x \mid x^2 + 1 = 0, \ x \text{ is a real number.}\}$

    (6) $\{x \mid x \text{ is an odd number bigger than } 10.\}$

#4 Find the value of $n(A) + n(B)$ for the following sets A and B :

(1) $A = \{x \mid x$ is a factor of 10.$\}$, $B = \{x \mid x$ is an even number less than 10.$\}$

(2) $A = \{x \mid 1 \leq x \leq 5,\ x$ is an odd number.$\}$, $B = \{0\}$

(3) $A = \{x \mid x$ is a natural number less than 1.$\}$, $B = \{x \mid 3 < x < 4,\ x$ is a natural number.$\}$

(4) $A = \{1, 2, 3, 4\}$, $B = \{2a + 1 \mid a \in A\}$

#5 Find the value of $a + b$ for the following sets A and B:

(1) $A = \{1, 2, a + 3\}$ and $B = \{2, 5, b + 1\}$, $A \subset B$ and $B \subset A$

(2) $A = \{3, 4, a + 1\}$ and $B = \{a + 2, b, 4\}$, $A \subset B$ and $B \subset A$

#6 Find the number of subsets and proper subsets for the following sets A :

(1) $A = \{x \mid x$ is a factor of 15.$\}$

(2) $A = \{x \mid x$ is an even number less than 8.$\}$

#7 $A = \{x \mid x \text{ is a factor of } 12.\}$ and $B = \{x \mid x \text{ is a factor of } 6.\}$

How many number of subsets which include all the elements of B are in the subsets of A?

#8 How many number of subsets which include the element a but not the elements b and c are in the subsets of $= \{a, b, c, d, e, f\}$?

#9 Find the number of a set A which satisfies the conditions for (1), (3), and (4). For (2), find the set A.

(1) $\{1\} \subset A \subset \{1, 2, 3\}$

(2) $\{2, 3\} \subset A \subset \{2, 3, 4, 5, 6\}$ and $n(A) = 3$

(3) $A \subset \{x \mid x \text{ is a natural number less than } 5.\}$ and A has at least one even number.

(4) $A \subset \{x \mid x \text{ is a factor of } 20.\}$ and $(1 \in A \text{ or } 2 \in A)$

#10 Find the value of $p + q$.

(1) The number of subsets of A is 64 and $n(A) = p$.

The number of proper subsets of B is 7 and $n(B) = q$.

(2) $A \subset \{x \mid 1 \leq x \leq p + q, \; x \text{ is a natural number.}\}$,

$(p \in A \text{ and } q \in A)$, and $n(A) = 32$, where $p + q > 2$

(3) $A \subset \{x \mid 1 \leq x \leq p + q, \ x \text{ is a natural number.}\}$,

    $(p \in A \text{ and } p + q \in A)$, $1 \notin A$, and $n(A) = 32$, where $p + q > 3$

#11 Find the intersection of the following sets

    (1) $A = \{x \mid x \text{ is a factor of } 6.\}$, $B = \{x \mid 1 \leq x \leq 10, \ x \text{ is an even number.}\}$

    (2) $A = \{1, 2, 3, 4, 5\}$, $B = \{x \mid x = a + 1, \ a \in A\}$

    (3) $A = \{x \mid x \text{ is a multiple of } 3.\}$, $B = \{x \mid x \text{ is a factor of } 12.\}$

    (4) $A = \{x \mid x \text{ is an even number.}\}$, $B = \{x \mid x \text{ is an odd number.}\}$

#12 Find the set A which satisfies the following conditions

    (1) $B = \{1, 2, 3\}$, $A \cup B = \{1, 2, 3, 4, 5\}$, $A \cap B = \emptyset$

(2) $B = \{1, 2, 3, 4\}$, $A \cup B = \{1, 2, 3, 4, 5, 6\}$, $A \cap B = \{1, 2\}$

(3) $B = \{1, 2, 3\}$, $A \cup B = \{1, 2, 3, 4, 5\}$, $n(A \cap B) = 2$

(4) $A = \{a, a + 1, a + 2\}$, $B = \{3, 4, 5, 6, 7\}$, $A \cap B = \{3, 4\}$

#13 Find the number of a set A with the following conditions

(1) $B = \{x \mid x \text{ is a factor of } 6.\}$, $C = \{x \mid x \text{ is a factor of } 18.\}$, $A \cap B = B$, $A \cup C = C$

(2) For a set B, $n(B) = 5$, $n(A \cap B) = 3$, and $n(A \cup B) = 10$

(3) For a set B, $A \cap B = \emptyset$, $n(B) = 7$, and $n(A \cup B) = 15$

(4) For a set B, $n(A \cup B) = 20$, $n(A \cap B) = 5$, and $n(B - A) = 8$

(5) For a fixed set $U = \{a, b, c, d, e, f, g\}$ and a set B,

$A - B = \{a, b\}$, $B - A = \{c, d\}$, and $(A \cup B)^C = \{f\}$

(6) For two sets B and C, $n(B \cup C) = 10$, $n(B \cap C) = 3$, $n(C) = 5$, and $A \subset B$, $A \cap C = \emptyset$

#14 Find the sets $A - B$ and $B - A$ for the following sets A and B.

(1) $A = \{1, 2, 3, 4, 5\}$, $B = \{x \mid x$ is a factor of 4 . $\}$

(2) $A = \{1, 2, 3, 4, 5, 6\}$, $B = \{x \mid 1 \leq x \leq 7,\ x$ is an odd number. $\}$

(3) For a fixed set $U = \{x \mid x$ is a factor of 20. $\}$,

$A \subset U$, $B \subset U$, $A \cap U = \{2, 5, 10\}$, $A \cap B = \{5, 10\}$, $(A \cup B)^C = \{1\}$

#15 Solve the following operations for any subsets A and B of the fixed set U.

(1) A^C

(2) $(A^C)^C$

(3) $U - A^C$

(8) $A - B$ when $A \cap B = A$

(4) $A - A^C$

(9) $A - B$ when $A \cap B = \emptyset$

(5) $(A \cup A^C) - (A \cap A^C)$

(10) $A \cap B$ when $A - B = A$

(6) $A \cap B^C$

(7) $A^C \cap B^C$

(11) $A^C - B^C$ when $A \subset B$

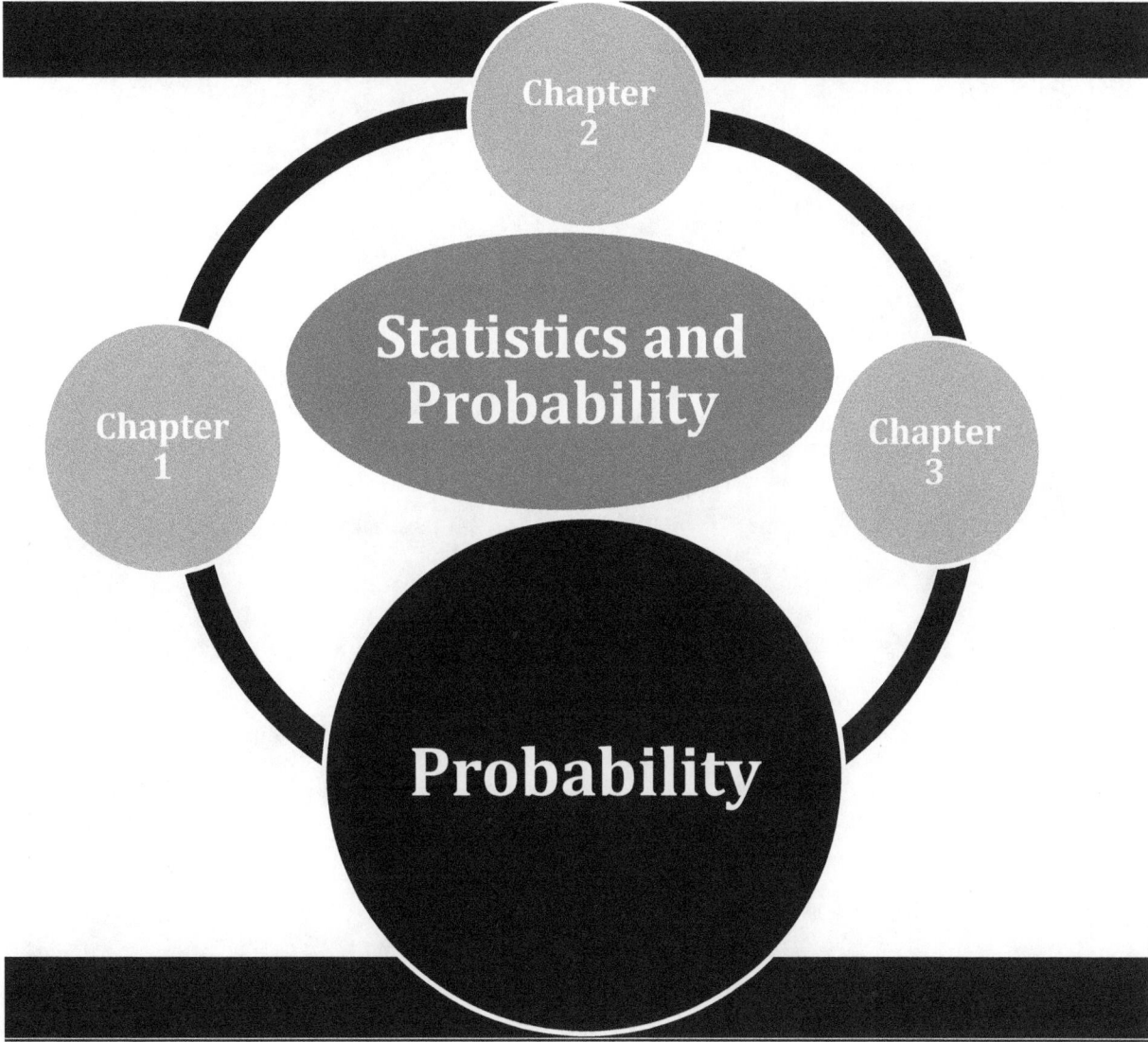

Chapter 2

Statistics and Probability

Chapter 1

Chapter 3

Probability

Chapter 4   Probability

CHAPTER 4

4-1 Probability

1. **Measuring Probability**

    (1) **Probability (Classical Definition)**

    (2) **Probability (Relative Frequency)**

2. **Sample Spaces and Events**

    (1) **Sample Spaces and Sample Points**

    (2) **Events**

        1) **Simple Event**

        2) **Compound Event**

3. **The Relationship between an Event and a Set**

    (1) **Union**

    (2) **Intersection**

    (3) **Complement**

    (4) **Mutually Exclusive(Disjoint) Events**

4. **Properties of Probability**

5. **Counting Outcomes**

    (1) **Tree Diagrams**

    (2) **The Multiplication Principle**

        1) **The Basic Rule of Counting**

        2) **The Generalized Basic Rule of Counting**

6. **Permutations**

4-2 Conditional Probability and Independence/Dependence

1. **Conditional Probability**

2. **Independence and Dependence**

Chapter 4. Probability

4-1 Probability

1. Measuring Probability

> If all of the outcomes have the same chance of occurrence, they are called *equally likely*.
> Example : When we toss a coin or throw a dice, we shall assume the possible outcomes are equally likely if not others mentioned.

(1) Probability (Classical Definition)

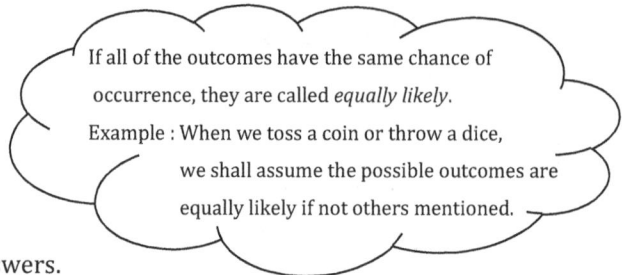

For some problems, there are no clear answers.

Probability deals with such randomness and uncertainty in mathematics.

Think about tossing a single coin.

If somebody asks, "What is the probability of obtaining heads? ", we usually answer that it is 50%, $\frac{1}{2}$, 0.5, or 1:2.

The probability is the ratio of the number of times a head obtains to the total number of outcomes (head, tail) in an experiment.

Suppose none of the outcomes (various results) occur at the same time.

If an experiment is repeated equally likely large number (N) of times, then the *probability* of an outcome (an even E), symbolized $P(E)$ is represented by

$$P(E) = \frac{N_E}{N} \text{ , where } N_E \text{ is the number of times an event } E \text{ occurs.}$$

> $$P(E) = \frac{\text{the number of times an event } E \text{ occurs}}{\text{the total number of possible outcomes}}$$

(2) Probability (Relative Frequency)

Tossing a single coin, the relative frequency of heads is quite changeable in two, three, or ten tosses of a coin. But after tossing a coin thousand times, it remains stable. If in many tosses of a coin, the proportion of heads observed is close to $\frac{1}{2}$, then we say that the probability of a head on a toss is $\frac{1}{2}$.

Thus, we say probability is long-term relative frequency.

If an experiment is performed large number (N) of times with outcomes (N_E) and the relative frequency $\frac{N_E}{N}$ approaches a limiting numerical number as the number N increases, then the probability of an outcome, $P(E)$, is represented by

$$P(E) = \lim_{n \to \infty} \frac{N_E}{N} \quad \text{, where } N_E \text{ is the number of times an event } E \text{ occurs.}$$

> Using the relative frequency, we can estimate the probability of particular outcome.
> The estimate of probability will be more accurate if more experiments are performed.

2. Sample Spaces and Events

(1) Sample Spaces and Sample Points

The *sample space* of an event is a set of all logical possible outcomes for a probability experiment and is denoted by $S = \{E_1, E_2, E_3, \cdots\cdots, E_n\}$.

Each element, $E_1, E_2, E_3, \cdots\cdots, E_n$ in a sample space is called a *sample point*.

Example

In a single flip of a coin, the sample space is {H, T}.

The sample space of flipping two coins is $S = \{(H, H), (H, T), (T, H), (T, T)\}$

Where $E_1 = \{(H, H)\}$: both coins are heads,

$\quad\quad E_2 = \{(H, T)\}$: the first coin is a head and the second is a tail,

$\quad\quad E_3 = \{(T, H)\}$: the first coin is a tail and the second is a head,

$\quad\quad E_4 = \{(T, T)\}$: both coins are tails.

(2) Events

Any collection (subset) of outcomes contained in the sample space for a probability experiment is called an *event*.

1) Simple Event

A *simple event* consists of exactly one outcome.

In this case, we usually write $E_1, E_2, E_3 \cdots\cdots$ instead of $\{E_1\}, \{E_2\}, \{E_3\}, \cdots\cdots$

2) Compound Event

A *compound event* consists of more than one event.

Example

For the sample space $S = \{(H, H), (H, T), (T, H), (T, T)\}$ of flipping two coins, there are 4 simple events,

$E_1 = \{(H, H)\}$, $E_2 = \{(H, T)\}$, $E_3 = \{(T, H)\}$, $E_4 = \{(T, T)\}$,

and some compound events including

$E = \{(H, H), (H, T), (T, H)\}$ (the event that at least one head shows) and $F = \{(H, T), (T, H), (T, T)\}$ (the event that at least one tail shows).

3. The Relationship between an Event and a Set

For any two events E and F of a sample space S,

(1) Union   The *union* of two events E and F, denoted by $E \cup F$, is the event consisting of all outcomes that are either in E or in F or in both E and F. The event $E \cup F$ will occur if either E or F occurs.

(2) Intersection   The *intersection* of two events E and F, denoted by $E \cap F$, is the event consisting of all outcomes that are in both E and F. The event $E \cap F$ will occur only if both E and F occur.

(3) Complement   The *complement* of an event E, denoted by E^C, is the set of all outcomes in sample space S that are not included in E.

Example

Consider rolling a die.

let $A = \{2, 4, 6\}$, $B = \{1, 3, 5\}$, $C = \{1, 2, 3, 4\}$. Then,

$A \cup B = \{1, 2, 3, 4, 5, 6\}$, $A \cap B = \emptyset$, $A \cap C = \{2, 4\}$, $A^C = \{1, 3, 5\}$, $(A \cup C)^C = \{5\}$.

(4) Mutually Exclusive (Disjoint) Events

Two events E and F are mutually exclusive if and only if $E \cap F = \emptyset$, where the null event, denoted by $\emptyset$, consists of no points.

That is, if two events E and F from the sample space have no outcomes in common, then E and F are *mutually exclusive* or *disjoint events*. $P(E \cap F) = 0$

Example

If $E = \{(H, H), (T, T)\}$ and $F = \{(H, T), (T, H)\}$, then $E \cap F = \emptyset$.

So, the two events E and F are mutually exclusive.

Note : Venn Diagrams

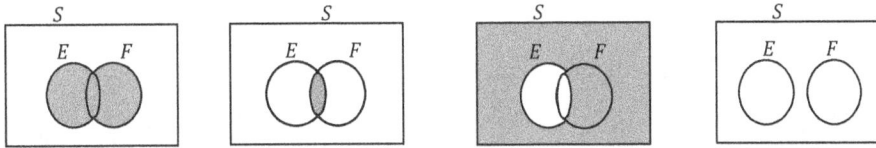

Shaded region is $E \cup F$.   Shaded region is $E \cap F$.   Shaded region is E^C.   E and F are mutually exclusive events.

4. Properties of Probability

Every random experiment has a sample space $S = \{E_1, E_2, E_3, \cdots\cdots, E_n\}$.

To each simple event E_i, $i = 1, 2, \cdots\cdots, n$, a probability $P(E_i)$, $i = 1, 2, \cdots\cdots, n$ is assigned.

(1) For each simple event E_i, $i = 1, 2, \cdots\cdots, n$, $0 \le P(E_i) \le 1$, $i = 1, 2, \cdots\cdots, n$ and $P(S) = 1$.

If $E_1, E_2, E_3, \cdots\cdots, E_n$ is a finite collection of mutually exclusive events, then

$$P(E_1 \cup E_2 \cup E_3 \cup \cdots\cdots \cup E_n) = P(E_1) + P(E_2) + P(E_3) + \cdots\cdots + P(E_n).$$

If $E \cap F = \emptyset$, then $P(E \cup F) = P(E) + P(F)$

(2) For any event E, $P(E) = 1 - P(E^C)$

($\because$ If E and E^C are mutually exclusive events, then $E \cup E^C = S$.

Since $P(S) = 1$, $1 = P(S) = P(E \cup E^C) = P(E) + P(E^C)$.

Therefore, $P(E) = 1 - P(E^C)$.)

$\circ \circ \bigcirc$ $P(E) + P(E^C) = 1$

(3) For any events E and F, if $E \cap F = \emptyset$, then $P(E \cap F) = 0$

($\because$ If an event $E \cap F$ contains no outcomes, then $(E \cap F)^C = S$.

Since $P(S) = 1$, $1 = P(S) = P((E \cap F)^C) = 1 - P(E \cap F)$.

Therefore, $P(E \cap F) = 0$.)

(4) For any events E and F, $P(E \cup F) = P(E) + P(F) - P(E \cap F)$

($\because$ Note that $E \cup F = E \cup (F \cap E^C)$ in a Venn Diagram.

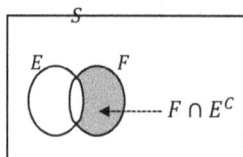

Since $E \cap (F \cap E^C) = \emptyset$, $P(E \cup F) = P(E) + P(F \cap E^C)$.

Note that $F = (E \cap F) \cup (F \cap E^C)$ in a Venn Diagram.

Since $(E \cap F) \cap (F \cap E^C) = \emptyset$, $P(F) = P(E \cap F) + P(F \cap E^C)$.

$\therefore P(F \cap E^C) = P(F) - P(E \cap F)$.

Therefore, $P(E \cup F) = P(E) + P(F) - P(E \cap F)$.)

5. Counting Outcomes

(1) Tree Diagrams

A tree diagram is used to display all possible outcomes of experiments with more than one event.

Example

The sample space of flipping two coins is $S = \{(H, H), (H, T), (T, H), (T, T)\}$.

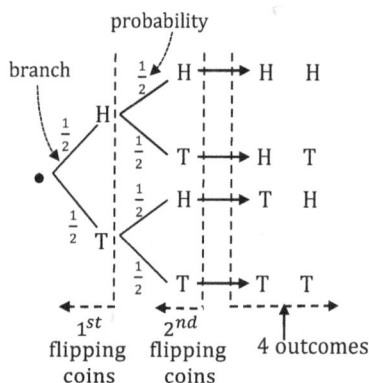

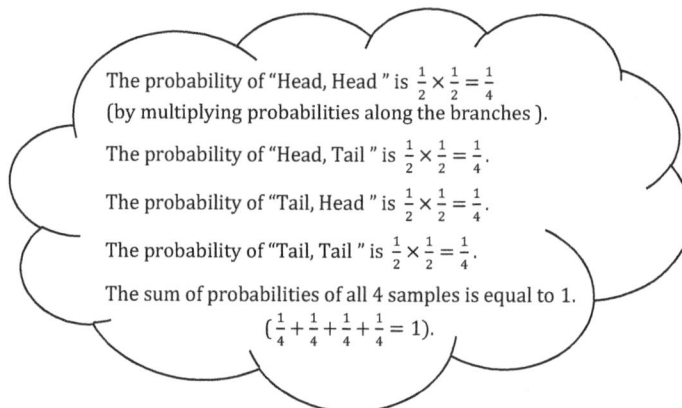

The probability of "Head, Head" is $\frac{1}{2} \times \frac{1}{2} = \frac{1}{4}$ (by multiplying probabilities along the branches).

The probability of "Head, Tail" is $\frac{1}{2} \times \frac{1}{2} = \frac{1}{4}$.

The probability of "Tail, Head" is $\frac{1}{2} \times \frac{1}{2} = \frac{1}{4}$.

The probability of "Tail, Tail" is $\frac{1}{2} \times \frac{1}{2} = \frac{1}{4}$.

The sum of probabilities of all 4 samples is equal to 1. ($\frac{1}{4} + \frac{1}{4} + \frac{1}{4} + \frac{1}{4} = 1$).

There are 2 branches for the 1st set of flipping coins and for each branch, there are 2 branches for the 2nd set of flipping coins. So, the total number of outcomes of the two experiments is $2 \times 2 = 4$.

(2) The Multiplication Principle

1) The Basic Rule of Counting

Given two experiments, if Experiment 1 has m possible outcomes, and if, for each outcome of Experiment 1, the Experiment 2 has n possible outcomes, there is a total of $m \times n$ possible outcomes of the two experiments.

2) The Generalized Basic Rule of Counting

Given n experiments, if the first experiment has m_1 possible outcomes, and if for each m_1 outcome the second experiment has m_2 possible outcomes, and if for each m_1 and m_2 outcomes the third experiment has m_3 possible outcomes, and if $\cdots\cdots$, then there is a total of $m_1 \times m_2 \times m_3 \cdots\cdots \times m_n$ possible outcomes of the n experiments.

Example

When flipping three coins,

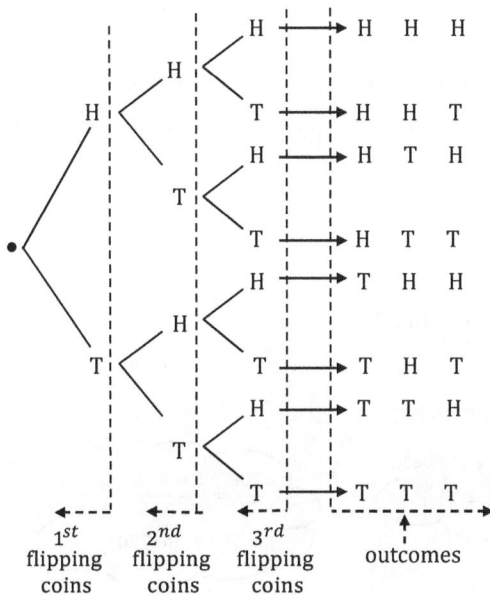

H	H	H
H	H	T
H	T	H
H	T	T
T	H	H
T	H	T
T	T	H
T	T	T

| 1st flipping coins | 2nd flipping coins | 3rd flipping coins | outcomes |

The total possible outcome of the 3 experiments is $2 \times 2 \times 2 = 2^3 = 8$.

Note :

Number of flipping coins	Total possible outcomes
1	2^1
2	2^2
3	2^3
$\vdots$	$\vdots$
m	2^m

When rolling n dice $\Rightarrow$ 6^n *outcomes*

When flipping m coins and rolling n dice $\Rightarrow$ $2^m \times 6^n$ *outcomes*

6. Permutations

Suppose we want to find the number of ways to arrange objects in which order does matter.

For letters A, B, and C, there are 6 different ordered arrangements, ABC, ACB, BAC, BCA, CAB, CBA.

Each arrangement is called a *permutation* of the three letters A, B, and C.

The number of possible permutation can be identified using the basic rule of counting.

We place the three possible letters for the first object. Each of these choices leaves two remaining choices for the second object, and only one choice remaining for the third object. This means there are $3 \cdot 2 \cdot 1 = 6$ possible different orders (permutations).

Now, consider n (a positive integer) objects.

In this case, there are $(n) \cdot (n-1) \cdot (n-2) \cdots (3) \cdot (2) \cdot (1)$ possible different permutations.

If a positive integer n is multiplied by all the preceding positive integers, the result is called n *factorial* and is denoted by $n!$ $(n! = nPn)$.

That is,

$$n \cdot (n-1) \cdot (n-2) \cdots 3 \cdot 2 \cdot 1 = n!$$

; the product of all positive integers less than or equal to n

Example $1! = 1$

$2! = 2 \cdot 1 = 2$

$3! = 3 \cdot 2 \cdot 1 = 6$

$4! = 4 \cdot 3 \cdot 2 \cdot 1 = 24$

0! is defined to be 1.
That is, $0! = 1$

$\dfrac{10!}{7!} = \dfrac{10 \cdot 9 \cdot 8 \cdots 2 \cdot 1}{7 \cdot 6 \cdot 5 \cdots 2 \cdot 1}$
$= 10 \cdot 9 \cdot 8 = 720$

The number of different ways of r objects that can be taken from a set of n obhects is symbolized by the notation nPr (the permutation of r objects taken from n objects) and is defined as the product of all positive integers r greatest factors of $n!$

That is, $nPr = \dfrac{n!}{(n-r)!} = \dfrac{n \cdot (n-1) \cdot (n-2) \cdots 3 \cdot 2 \cdot 1}{(n-r) \cdot (n-r-1) \cdot (n-r-2) \cdots 3 \cdot 2 \cdot 1} = n \cdot (n-1) \cdot (n-2) \cdots (n-r+1)$

Example How many different ways of 3 letters could be selected from a total of 10 letters without replacement? $10P3 = 10 \cdot 9 \cdot 8 = 720$ different ways

We can use permutations if all objects are chosen from the same group, if no object may be used more than once, and if the order of arrangement matters.

For example, consider the different letter arrangements formed by the letters MIRROR.

In this case, there are

$\dfrac{6!}{1! \cdot 1! \cdot 3! \cdot 1!} = \dfrac{6 \cdot 5 \cdot 4 \cdot 3!}{1! \cdot 1! \cdot 3! \cdot 1!} = 6 \cdot 5 \cdot 4 = 120$ possible letter arrangements.

In the case of PEPPER, there are

$$\frac{6!}{3!\cdot 2!} = \frac{6\cdot 5\cdot 4\cdot 3!}{3!\cdot 2} = 60 \text{ possible letter arrangements.}$$

> If in a set of n objects, r objects are identical, there are $\dfrac{n!}{r!}$ different ways.

Example How many different ways, each consisting of 7 flags hung in a line, can be arranged from a set of 3 red flags, 2 white flags, and 2 blue flags if all flags of the same color are identical ?

$$\frac{7!}{3!2!2!} = \frac{7\cdot 6\cdot 5\cdot 4\cdot 3\cdot 2\cdot 1}{(3\cdot 2\cdot 1)(2\cdot 1)(2\cdot 1)} = 7\cdot 5\cdot 3\cdot 2\cdot 1 = 210 \text{ different ways}$$

Note : *The number of different ways of r objects that can be taken from a set of n objects is symbolized by the notation nCr (the combination of r objects taken from n objects) for $r \le n$.*

We define

$$nCr = \frac{nPr}{r!} = \frac{n!}{(n-r)!\,r!} \quad for \ r \le n$$

> Subgroups of r objects (which are not arranged in order) are selected from a set of n objects.

The number of possible combinations of n objects taken r at a time.

Example

A group of 5 is to be formed from a class of 20 students. How many different groups are possible?

There are $20C5 = \dfrac{20P5}{5!} = \dfrac{20!}{(20-5)!\,5!} = \dfrac{20\cdot 19\cdot 18\cdot 17\cdot 16\cdot 15!}{15!(5\cdot 4\cdot 3\cdot 2\cdot 1)} = 19\cdot 3\cdot 17\cdot 16 = 15{,}504 \ possible \ groups.$

> Since $nCn-r = \dfrac{n!}{(n-n+r)!(n-r)!} = \dfrac{n!}{r!(n-r)!}$,
>
> $nCr = nCn-r$.
>
> Note that $nCn = nCo = 1$ and
>
> $nC1 = n$

> If the order does not matter, it is a combination.
> If the order does matter, it is a permuattion.

4-2 Conditional Probability and Independence/Dependence

1. Conditional Probability

For any two events E and F with $P(F) > 0$,

the conditional probability of E given that F has occurred is defined by

$$\boxed{P(E\backslash F) = \frac{P(E\cap F)}{P(F)}}$$

Note : $\boxed{P(E \cap F) = P(E\backslash F)\cdot P(F)}$

> Generally,
> $P(E\backslash F) \ne P(E)$
> $P(F\backslash E) \ne P(F)$

2. Independence and Dependence

If the knowledge that F has occurred does not change the probability that E occurs or has occurred, then $P(E \backslash F) = P(E)$. In this case, we say that E is *independent* of F.

Since $P(E \backslash F) = \frac{P(E \cap F)}{P(F)}$,

$$\boxed{E \text{ is independent of } F \text{ if } P(E \cap F) = P(E) \cdot P(F)}$$

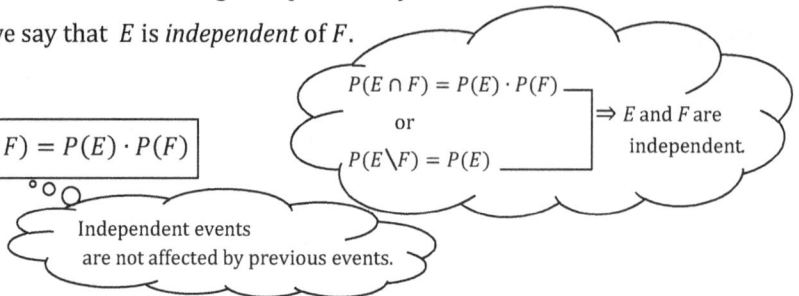

$$\boxed{\begin{array}{c} P(E \cap F) = P(E) \cdot P(F) \\ \text{or} \\ P(E \backslash F) = P(E) \end{array}} \Rightarrow \begin{array}{c} E \text{ and } F \text{ are} \\ \text{independent.} \end{array}$$

Independent events are not affected by previous events.

If two events E and F are not independent: that is, the outcome of one event does affect the outcome of the other, then we say that E and F are *dependent*.

Note :   ① *To find the probability of two independent events both occurring,*

multiply the probability of the first event (E) by the probability of the second event (F).

P *(E and F)* $= P(E) \cdot P(F)$

② *To find the probability of two dependent events both occurring,*

multiply the probability of the first event (E) by the probability of the second event (F) after E occurs .

P *(E and F)* $= P(E) \cdot \underline{P(F \text{ following } E)}$

the conditional probability of F given that E has occurred

Example 1

Consider flipping a coin and rolling a die.

The probability of obtaining heads in one coin flip is $\frac{1}{2}$ and the probability of obtaining a number 3 in one die roll is $\frac{1}{6}$. In this case, the occurrence of one event does not affect the occurrence of the other.

So, the two events are independent. Therefore, the probability of both occurring is $\frac{1}{2} \cdot \frac{1}{6} = \frac{1}{12}$.

Example 2

Consider rolling a die.

Define the events $E = \{1, 2, 3\}$, $F = \{2, 4, 6\}$, and $G = \{1, 3, 5, 6\}$.

Then, $P(E) = \frac{3}{6} = \frac{1}{2}$, $P(E \backslash F) = \frac{P(E \cap F)}{P(F)} = \frac{\frac{1}{6}}{\frac{3}{6}} = \frac{1}{3}$, and $P(E \backslash G) = \frac{P(E \cap G)}{P(G)} = \frac{\frac{2}{6}}{\frac{4}{6}} = \frac{1}{2}$.

Since $P(E) = P(E \backslash G) = \frac{1}{2}$, events E and G are independent.

Since $P(E) \neq P(E \backslash F)$, events E and F are dependent.

Exercises

#1 Find the sample space for the following

   (1) A spin of a spinner marked 0, 1, 2, 3

   (2) A toss of one coin and a spin of a spinner marked 0, 1, 2, 3

   (3) Two coins tossed once

   (4) A pair of dice tossed once

   (5) A toss of one coin and a toss of an ordinary die.

#2 Think about tossing one coin and one ordinary die. Find the probability of the event

   (1) E : two heads occur.

   (2) F : a head and an even number occurs.

   (3) G : an odd or even number occurs.

#3 Find the probabilities for the indicated sample spaces derived from the random experiment of drawing one card from a full deck.

   (1) $S = \{red , black\}$

   (2) $S = \{ace\ or\ picture\ card , otherwise\}$

#4   One ball is drawn from a bag containing three red balls marked 1, 2, 3 ; four blue balls marked 1, 2, 3, 4 ; and two yellow balls marked 1, 2. Find the probability for the indicated sample spaces from this random experiment.

    (1) $S = \{\text{red}, \text{ blue}, \text{ yellow}\}$

    (2) $S = \{\text{even}, \text{odd}\}$

#5   Three coins are tossed at once. Find $P(E \cap F)$.

    (1) Let E be the event "coins match" and let F be the event "not more than one head".

    (2) Let E be the event "coins match" and let F be the event "not more than three heads".

    (3) Let E be the event "coins match" and let F be the event "at least two heads".

    (4) Let E be the event "coins match" and let F be the event "at least one head".

    (5) Let E be the event "head on first toss" and let F be the event "tail on second toss".

#6 For the following experiments, one toss is made. Find the probabilities indicated.

(1) Two coins, E : "at most one head", F : "no tails". Find $P(E \cup F)$

(2) Three coins, E : "at least two heads", F : "only one tail". Find $P(E \cup F)$.

(3) Two dice, $E = \{(1, 2)\}$, $F = \{(3, 4)\}$. Find $P(E \cup F)$.

(4) Two dice, E : "The sum of marked numbers is 6". Find $P(E)$.

(5) Two dice, E : "sum ≤ 10". Find $P(E)$.

#7 Find the number of different arrangements (permutations) for the following events

(1) Scheduling seven different classes in seven periods.

(2) Creating the batting order for a baseball team consisting of 9 players.

#8   A coin is flipped twice. What is the conditional probability that both coins are heads, given that the first coin is a head?

#9   A bag contains 5 red, 7 white, and 10 black balls. A ball is chosen at random from the bag, and it is noted that it is not one of the white balls. What is the conditional probability that it is red?

#10   A box contains 10 apples and 15 pears. The fruits to be chosen are selected at random. Find the probability that

(1) The first two fruits chosen are apples.

(2) The second fruit chosen is an apple.

(3) Given that the second fruit chosen is an apple, the first fruit chosen is also an apple.

#11 Think about tossing a coin and an ordinary dice. Let E be the event " H on coin " and let F be the event " 2 on dice ". Find $P(E \cup F), P(E \cap F)$, and determine if E and F are dependent or independent.

#12 Think about tossing a nickel and a dime. Let E be the event " coins match ", let F be the event " nickel falls on heads ", and let G be the event " at least one head shows ". Determine which events are independent.

Solutions Manual

CH. 1 ~ CH. 3

Algebra

Part III. Functions

Solutions for Chapter 1

#1. The domain of a function $f(x) = -2x + 3$ is $\{0,\ 1,\ 2,\ 3\}$. Find the range of $f(x)$.

$x = 0 \Rightarrow f(0) = -2 \cdot 0 + 3 = 3$

$x = 1 \Rightarrow f(1) = -2 \cdot 1 + 3 = 1$

$x = 2 \Rightarrow f(2) = -2 \cdot 2 + 3 = -1$

$x = 3 \Rightarrow f(3) = -2 \cdot 3 + 3 = -3$

∴ The range of $f(x)$ is $\{-3, -1,\ 1,\ 3\}$.

#2. The range of a function $f(x) = 2x$ is $\{-8,\ 0,\ 4,\ 8\}$. Find the domain of $f(x)$.

$2x = -8 \Rightarrow x = -4$

$2x = 0 \Rightarrow x = 0$

$2x = 4 \Rightarrow x = 2$

$2x = 8 \Rightarrow x = 4$

∴ The domain of $f(x)$ is $\{-4,\ 0,\ 2,\ 4\}$.

#3. Find the domain and range of the equation $y = |x| - 3$.

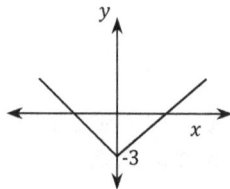

∴ The domain of $f(x)$ is the set of all real numbers and the range of $f(x)$ is the set of all real numbers

that are greater than or equal to -3 (∵ Since $|x| \geq 0$, $|x| - 3 \geq -3$).

#4. The range of a function $g(x) = ax$ is $\{-2, \ 0, \ 2\}$ when $g(2) = -1$. Find the domain of the function.

Since $g(2) = 2a = -1$, $a = -\frac{1}{2}$ $\quad \therefore g(x) = -\frac{1}{2}x$

Since $-\frac{1}{2}x = -2 \Rightarrow x = 4$, $-\frac{1}{2}x = 0 \Rightarrow x = 0$, and $-\frac{1}{2}x = 2 \Rightarrow x = -4$,

the domain of the function is $\{-4, \ 0, \ 4\}$.

#5. $A = \{(-3, 1), (-2, 2), (-2, 3), (-1, 4), (0, 5), (1, 5)\}$ is the set of ordered pairs.

Is this relationship a function?

No ($\because$ -2 is assigned to two values, 2 and 3).

#6. For a function $f(x) = ax$, $f(3) = -4$. Find the value of $f(9)$.

Since $f(3) = 3a = -4$, $a = -\frac{4}{3}$ $\quad \therefore f(x) = -\frac{4}{3}x$

So, $f(9) = -\frac{4}{3} \cdot 9 = -12$

#7. Find the value of $f(3) - f(2) + f(4)$ for the function $f(x) = \frac{3}{x}$.

Since $f(3) = \frac{3}{3} = 1$, $f(2) = \frac{3}{2}$, and $f(4) = \frac{3}{4}$, $f(3) - f(2) + f(4) = 1 - \frac{3}{2} + \frac{3}{4} = \frac{4-6+3}{4} = \frac{1}{4}$

#8. For the two functions $f(x) = ax + 2$ and $g(x) = \frac{b}{x} - 2$, $f(1) = g(-1) = 3$.

Find the value of $a + b$.

Since $f(1) = g(-1) = 3$, $a + 2 = -b - 2 = 3$

$\therefore a = 1, \ b = -5$

Therefore, $a + b = -4$

#9. For the two functions $f(x) = \frac{a}{x} + 2$ **and** $g(x) = -\frac{3}{x} + 5$, $\quad 3f(-2) = 2g(-3)$.

Find the value of b **which satisfies** $f(b) = g(b)$.

Since $3f(-2) = 2g(-3)$, $\quad 3\left(\frac{a}{-2} + 2\right) = 2\left(-\frac{3}{-3} + 5\right)$; $\quad -\frac{3a}{2} + 6 = 12$; $\quad -\frac{3a}{2} = 6$; $\quad a = -4$

Since $f(b) = g(b)$, $\quad \frac{a}{b} + 2 = -\frac{3}{b} + 5$; $\quad \frac{-4}{b} + 2 = -\frac{3}{b} + 5$; $\quad \frac{1}{b} = -3$; $\quad b = -\frac{1}{3}$

$\therefore$ The value of b is $-\frac{1}{3}$.

#10. Identify functions.

(1)

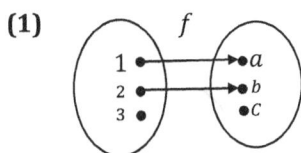

(2)

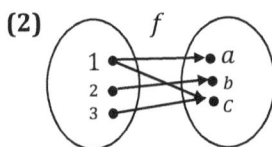

(1) 3 is not assigned to any number. So, it's not a function.

(2) 1 is assigned to two values. So, it's not a function.

(3)

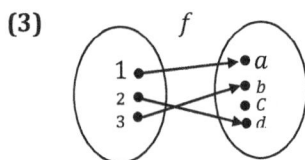

(4)

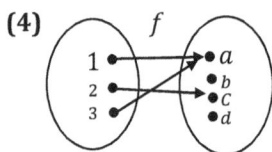

(3) It's a function.   (4) It's a function.

(5)

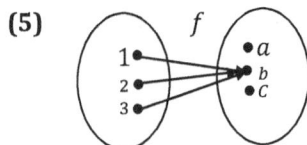

(6)

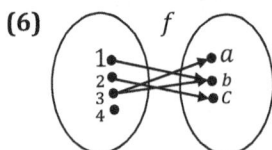

(5) It's a function.

(6) 3 is assigned to two values and also, 4 is not assigned to any number. So, it's not a function.

#11. For the function $f(3x - 2) = 2x - a$, $f(4) = 3$. **Find the value of** $f(1)$.

To find $f(4)$, $3x - 2 = 4$; $3x = 6$; $x = 2$

Since $f(4) = 3$, $f(4) = f(3 \cdot 2 - 2) = 2 \cdot 2 - a = 3$ $\therefore$ $a = 1$

$\therefore f(3x - 2) = 2x - 1$

Therefore, $f(1) = f(3 \cdot 1 - 2) = 2 \cdot 1 - 1 = 1$

#12. For the two functions $f(x) = 2ax$ **and** $g(x) = \frac{2}{x} - 1$, $g(f(2)) = 3$. **Find the value of** a.

Since $f(2) = 2a \cdot 2 = 4a$, $g(f(2)) = g(4a) = \frac{2}{4a} - 1 = \frac{1}{2a} - 1 = 3$

$\therefore \frac{1}{2a} = 4$

Therefore, $a = \frac{1}{8}$

#13. Plot the following ordered pairs on the graph.

(1) $A(2, 3)$
(2) $B(-2, 3)$
(3) $C(2, -3)$
(4) $D(-5, 5)$
(5) $E(0, 5)$
(6) $F(4, 0)$
(7) $G(-3, 0)$
(8) $H(0, -7)$

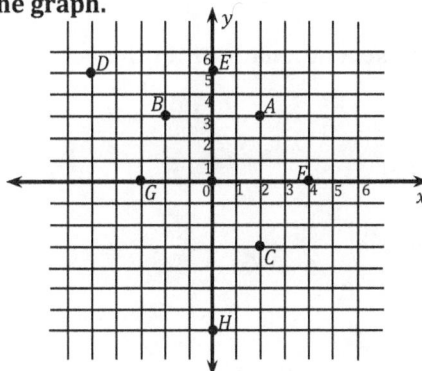

#14. Find the coordinates for each point on the graph.

(1) $A = A(2, 2)$
(2) $B = B(5, 0)$
(3) $C = C(1, -4)$
(4) $D = D(-3, -2)$
(5) $E = E(-3, 2)$
(6) $F = F(0, 6)$

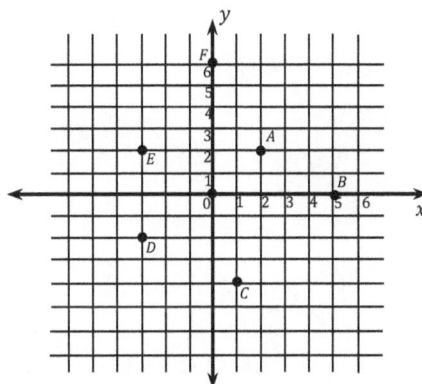

#15. Two points $P(a+2, 4-2a)$ and $Q(2-2b, 3b+1)$ are on the x-axis and y-axis, respectively.

Find the value of $a + b$.

$4 - 2a = 0$ and $2 - 2b = 0$

$\therefore a = 2$ and $b = 1$

Therefore, $a + b = 3$

#16. Find the length of the segment between the two points.

(1) $A(1, 2)$ and $B(1, -2)$; $AB = 4$

(2) $C(0, 3)$ and $D(-3, 3)$; $CD = 3$

(3) $P(-3, 0)$ and $Q(5, 0)$; $PQ = 8$

(4) $S(-5, 0)$ and $T(-5, -6)$; $ST = 6$

#17. A point (a, b) is in the second quadrant of the coordinate plane. Name the quadrant containing the following points

Note that $a < 0$ and $b > 0$

(1) $(a, -b)$ III

(2) $(-b, a)$ III

(3) (b, a) IV

(4) $(-a, -b)$ IV

(5) $(-a, b)$ I

(6) $(-b, -a)$ II

(7) (ab, a^2) II ($\because ab < 0, a^2 > 0$)

(8) $(-a, -ab)$ I ($\because -a > 0, -ab > 0$)

#18. Point B is reflected through the origin to point $A(3,4)$. Point C is obtained by reflecting point B across the y-axis. Find the area of a triangle $\triangle ABC$.

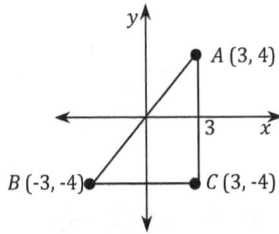

Since the height AC is 8 and the length BC is 6, the area of a triangle $\triangle ABC$ is $\frac{1}{2} \cdot 6 \cdot 8 = 24$.

#19. Point $C(4,b)$ is the midpoint of Points $A(-2,3)$ and $B(a,9)$. Find the value of $a-b$.

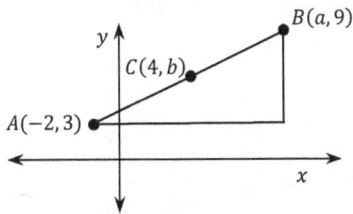

Since $\frac{a-2}{2} = 4$ and $\frac{9+3}{2} = b$, $a = 10$ and $b = 6$.

Therefore, $a - b = 10 - 6 = 4$

#20. Which graphs are functions? (3), (5), and (7)

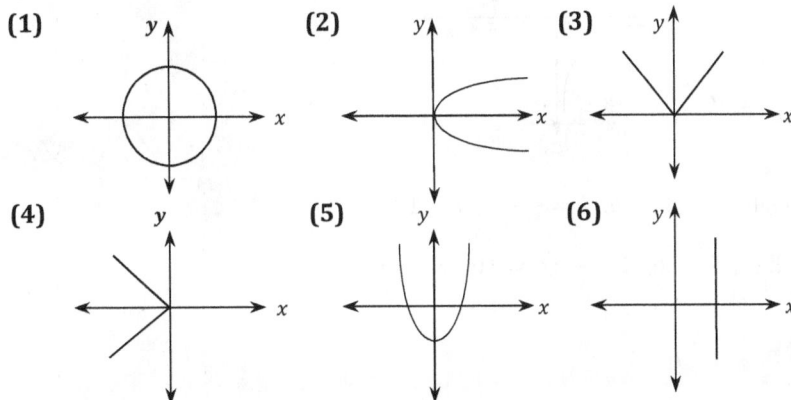

(1)

(2)

(3)

(4)

(5)

(6)

(7)

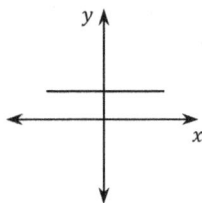

(8)

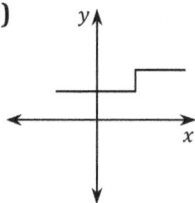

#21. Identify the function of the form $y = ax$ which passes through the origin and $(3, -4)$.

Since $y = ax$, $-4 = 3a$

$\therefore a = -\dfrac{4}{3}$

Therefore, $y = -\dfrac{4}{3}x$

#22. Identify the functions of the form $y = ax$ or $y = \dfrac{a}{x}$ for the following graphs

(1)

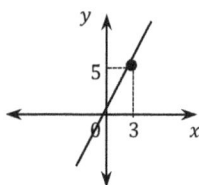

(2)

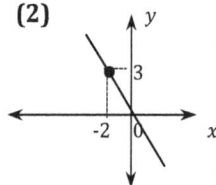

(1) $y = ax$ passes through $(3, \ 5)$. So, $5 = 3a$; $a = \dfrac{5}{3}$ $\therefore y = \dfrac{5}{3}x$

(2) $y = ax$ passes through $(-2, \ 3)$. So, $3 = -2a$; $a = -\dfrac{3}{2}$ $\therefore y = -\dfrac{3}{2}x$

(3)

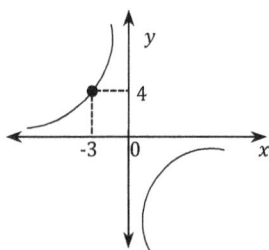

(4)

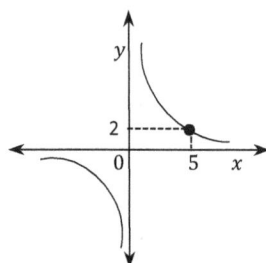

(3) $y = -\dfrac{a}{x}$ passes through $(-3, \ 4)$. So, $4 = -\dfrac{a}{-3}$; $a = 12$ $\therefore y = -\dfrac{12}{x}$

(4) $y = \dfrac{a}{x}$ passes through $(5, \ 2)$. So, $2 = \dfrac{a}{5}$; $a = 10$ $\therefore y = \dfrac{10}{x}$

#23. Find the functions for the data in the tables below.

(1) $y = \frac{4}{x}$

x	-4	-2	-1	1	2	4
y	-1	-2	-4	4	2	1

(2) $y = 2x + 3$

x	-2	-1	0	1	2	3
y	-1	1	3	5	7	9

(3) $y = \frac{12}{x}$

x	1	2	3	4	6	12
y	12	6	4	3	2	1

(4) $y = -\frac{1}{2}x$

x	-4	-2	0	2	4	6
y	2	1	0	-1	-2	-3

#24. The function $f(x) = -\frac{3}{2}x$ passes through a point $(a+1, 2a-3)$. Find the value of a.

$2a - 3 = -\frac{3}{2}(a+1)$; $4a - 6 = -3a - 3$; $7a = 3$ $\therefore a = \frac{3}{7}$

#25. The function $y = ax$ passes through a point $(3, -15)$ and $(b, 10)$. Find the value of $a - b$.

Since $-15 = 3a$, $a = -5$

Since $10 = ab = -5b$, $b = -2$

Therefore, $a - b = -5 - (-2) = -3$

#26. For any constants a and b, the function $f(x) = \frac{2a}{x}$ passes through the points $(-2, 8)$ and $(4, b)$.

Find the value of $a + b$.

$8 = \frac{2a}{-2}$; $a = -8$ $\therefore f(x) = \frac{-16}{x}$

$b = \frac{-16}{4}$; $b = -4$

Therefore, $a + b = -12$

#27. For any constants $a, b,$ and c, the function $f(x) = \frac{a}{x}$ passes through the points $(b, 1)$, $(1, c)$, and $(3, -1)$. Find the value of $a + b + c$.

$-1 = \frac{a}{3}$; $a = -3$ $\therefore f(x) = \frac{-3}{x}$

Since $1 = \frac{-3}{b}$, $b = -3$

Since $c = \frac{-3}{1}$, $c = -3$

Therefore, $a + b + c = -3 + (-3) + (-3) = -9$

#28. Two functions $f(x) = ax$ and $g(x) = \frac{b}{x}$ meet at the points $(3, 9)$ and $(-3, c)$. Find the value of $a + b + c$.

Since $f(x) = ax$, $9 = 3a$; $a = 3$

Since $g(x) = \frac{b}{x}$, $9 = \frac{b}{3}$; $b = 27$

Since $(-3, c)$ is on $f(x) = 3x$, $c = 3(-3) = -9$

Therefore, $a + b + c = 3 + 27 + (-9) = 21$

#29. Two functions $y = -ax$ and $y = -\frac{2}{x}$ meet at Point $A(b, 8)$. Find the value of ab.

Since $y = -\frac{2}{x}$, $8 = -\frac{2}{b}$ $\therefore b = -\frac{1}{4}$

Since $y = -ax$, $8 = -ab = \frac{1}{4}a$ $\therefore a = 32$

Therefore, $ab = 32 \cdot \left(-\frac{1}{4}\right) = -8$

#30. Find the function of the form $y = ax$ or $y = \frac{a}{x}$ for the graph.

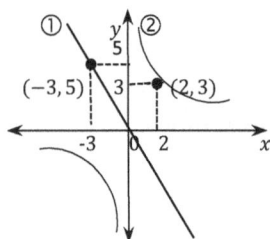

① $y = ax$; $5 = -3a$; $a = -\frac{5}{3}$ $\therefore y = -\frac{5}{3}x$

② $y = \frac{a}{x}$; $3 = \frac{a}{2}$; $a = 6$ $\therefore y = \frac{6}{x}$

#31. The function $y = 3x$ passes through the two points, origin and A. The area of the triangle $\triangle OAB$ is 54. Find the coordinate of Point A.

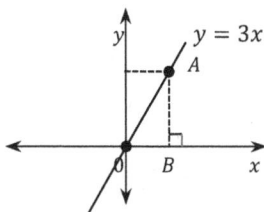

$A(a, 3a)$, $B(a, 0)$

∴ The area of the triangle $\triangle OAB$ is $\frac{1}{2} \cdot a \cdot 3a = 54$; $3a^2 = 108$; $a^2 = 36$ ∴ $a = 6$ $(\because a > 0)$

Therefore, $A(a, 3a) = A(6, 18)$

#32. Two points $P(3, a)$ and $Q(3, b)$ are on the graph $y = 3x$ and $y = -x$, respectively.

    Find the area of the triangle $\triangle OPQ$.

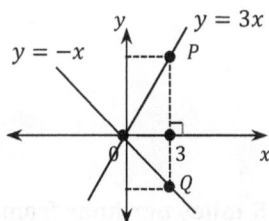

$P(3, a) = P(3, 9)$ and $Q(3, b) = Q(3, -3)$,

Since the height of the triangle $\triangle OPQ$ is 3 and the length of it is $PQ = 12$,

the area of the triangle $\triangle OPQ$ is $\frac{1}{2} \cdot 12 \cdot 3 = 18$ square units.

#33. Richard rides his bike from home to a park 5 miles away at a speed of x miles per hour for y hours. Find the relationship between x and y.

    Note Distance = speed × time (in hour)

∴ $5 = xy$

Therefore, $y = \frac{5}{x}$

#34. Which one is not a function? (3)

      (1) The sum of two variables x and y is 5.

      (2) The variable y is half of the variable x.

      (3) The perimeter (y inch) of a rectangle with one side of length (x inch).

      (4) 10 miles at a speed of x miles per hour for y minutes.

#35. A building needs to be painted. It takes 30 hours for 5 workers to finish the job. If the job has to be finished in 6 hours, how many workers are needed?

    Let x be the number of workers needed and y be the total time to finish the job.

    Since it takes 30 hours for 5 workers to finish the job, $xy = 30 \cdot 5$

    $\therefore y = \frac{150}{x}$

    Since the job has to be finished in 6 hours, $6 = \frac{150}{x}$; $x = \frac{150}{6} = 25$

    $\therefore$ 25 workers are needed to finish the job in 6 hours.

#36. Nichole wants to make a vegetable garden with an area of 200 square feet. Find the relation between the length (x feet) and width (y feet).

    Since $xy = 200$, $y = \frac{200}{x}$

#37. The distance from A to B is 10 miles. Nichole drives at a speed of 35 miles per hour from A to B, and Richard drives at a speed of 25 miles per hour from B to A at the same time. How long will it take before they meet each other?

    Let t be the time (in hour).

    Then, $35t + 25t = 10$; $60t = 10$; $t = \frac{1}{6}$

    $\therefore \frac{1}{6}$ hour (10 minutes)

#38. Richard drives to a post office at a speed of 50 miles per hour. 5 minutes later, Nichole drives to the post office at a speed of 60 miles per hour. How long will it take before Nichole meets Richard?

    Let t be the time Nichole drives until they meet.

    When they meet each other, they travel the same distance. Thus, $50\left(t + \frac{1}{12}\right) = 60t$.

    $\therefore 10t = \frac{50}{12}$; $t = \frac{50}{12} \cdot \frac{1}{10} = \frac{5}{12}$

    $\therefore \frac{5}{12}$ hour (25 minutes)

#39. Richard drives to school at a speed of 40 miles per hour and returns back home at a speed of 30 miles per hour. Coming home, it takes him 10 more minutes than going school. How far is it from Richard's home to school?

Let t be the time it takes Richard to get to school at 40 mph.

$40t = 30(t + \frac{10}{60})$; $10t = 5$; $t = \frac{1}{2}$

Therefore, the distance is $40t = 40 \cdot \frac{1}{2} = 20$. Therefore, the distance is 20 miles.

#40. Nichole rides her bike halfway to school at 20 mph. She drives her car the rest of the way at 40 mph. Find Nichole's average speed to school.

Let t_1 be the time for riding her bike, t_2 be the time for driving her car, and D be the total distance.

Then, $\frac{1}{2}D = 20t_1$ and $\frac{1}{2}D = 40t_2$

$\therefore t_1 = \frac{1}{40}D$ and $t_2 = \frac{1}{80}D$

Since $t_1 + t_2 = \frac{3}{80}D$, the average speed is $V = \frac{D}{\frac{3}{80}D} = \frac{80}{3} = 26\frac{2}{3}$

Therefore, the average speed to school is $26\frac{2}{3}$ miles per hour.

#41. x ounces of a y% salt solution contains 3 ounces of salt.
Find the relationship between x and y.

$x \cdot \frac{y}{100} = 3$ $\therefore y = \frac{300}{x}$

#42. Nichole wants to buy some books at a bookstore which are all the same price. If she buys 3 books, then she will be $2.50 short. If she buys 2 books, then she will have $5.00 left over. How much money does she have?

Let x be the price of the books.

Then $3x - 2.50 = 2x + 5$; $x = 7.50$

$\therefore 2 \times 7.50 + 5 = 20$

Therefore, she has $20.

#43. 3 machines can do 5 jobs in 4 hours. How many hours will it take for 4 machines to do 6 jobs?

Since 1 machine can do $\frac{5}{3}$ jobs in 4 hours, 1 machine can do $\frac{\frac{5}{3}}{4} = \frac{5}{12}$ job in 1 hour.

So, 4 machines can do $\frac{5}{12} \cdot 4 = \frac{5}{3}$ jobs in 1 hour.

$$\therefore \frac{5}{3} : 1 = 6 : x \quad \therefore \frac{5}{3}x = 6 \; ; \; x = 6 \cdot \frac{3}{5} = \frac{18}{5}$$

Therefore, $\frac{18}{5}$ hours (3 hours 36 minutes)

OR By the formula : $W = mtr$ (W work done by m in time t at a constant rate r),

$$5 = 3 \cdot 4 \cdot r \quad \therefore \; r = \frac{5}{12}$$

So, $6 = 4 \cdot x \cdot \frac{5}{12}$; $x = \frac{6 \cdot 12}{4 \cdot 5} = \frac{18}{5} = 3\frac{3}{5}$ hours (3 hours 36 minutes)

Solutions for Chapter 2

#1 Identify linear functions. Mark O for a linear function or $\times$ for a non-linear function.

(1) $y = 2 + x$; O

(2) $2x + y = 3$; O

(3) $y = \frac{2}{x} + 1$; $\times$

(4) $x^2 + y^2 = 1$; $\times$

(5) $y = x^2 + 2x + 1$; $\times$

(6) $x + y = 0$; O

(7) $y = 2(x + 1) - 2x$; $\times$

(8) $y = x^2 - (x + 2)^2$; O

(9) $y = 1$; $\times$

(10) $xy = 2$; $\times$

#2 Find the following values for the linear function $f(x) = 2x + 3$

(1) $f(0)$; $f(0) = 2 \cdot 0 + 3 = 3$

(2) $f(1) + f(-1)$; $f(1) = 2 \cdot 1 + 3 = 5$ and $f(-1) = -2 + 3 = 1$ $\therefore f(1) + f(-1) = 6$

(3) $\frac{1}{2}f(-2) \cdot f\left(\frac{1}{2}\right)$

Since $f(-2) = -4 + 3 = -1$, $\frac{1}{2}f(-2) = -\frac{1}{2}$

Since $f\left(\frac{1}{2}\right) = 1 + 3 = 4$, $\frac{1}{2}f(-2) \cdot f\left(\frac{1}{2}\right) = -\frac{1}{2} \cdot 4 = -2$

(4) $f(f(2))$; Since $f(2) = 2 \cdot 2 + 3 = 7$, $f(f(2)) = f(7) = 2 \cdot 7 + 3 = 17$

#3 Find the following values for the given linear functions with a condition

(1) $f(1)$ for $f(x) = 2ax + 1$ with $f(-1) = 3$

$f(-1) = -2a + 1 = 3$; $2a = -2$; $a = -1$ $\therefore f(x) = -2x + 1$

Therefore, $f(1) = -2 + 1 = -1$

(2) $\dfrac{a}{2}$ for $f(x) = \dfrac{1}{2}x + 5$ with $f\left(\dfrac{a}{2}\right) = -a$

$f\left(\dfrac{a}{2}\right) = \dfrac{1}{2}\cdot\dfrac{a}{2} + 5 = \dfrac{a}{4} + 5 = -a$; $\dfrac{5a}{4} = -5$; $a = -4$

$\therefore \dfrac{a}{2} = -2$

(3) $a + b$ for $f(x) = 3ax - 2$ with $f(-1) = 4$ and $f(b) = 1$

$\qquad f(-1) = -3a - 2 = 4$; $3a = -6$; $a = -2$ $\therefore f(x) = -6x - 2$

$\qquad$ So, $f(b) = -6b - 2 = 1$; $b = -\dfrac{1}{2}$

$\qquad \therefore a + b = -\dfrac{5}{2}$

(4) $a + \dfrac{1}{a}$ when $f(x) = 3x - 1$ passes through the point $(a, a + 3)$.

$\qquad a + 3 = 3a - 1$; $2a = 4$; $a = 2$ $\therefore a + \dfrac{1}{a} = \dfrac{5}{2}$

(5) $a - b$ when $f(x) = ax + 2$ passes through both point $(1, 3)$ and point $(2, b)$.

$\qquad$ Substitute $(1, 3)$ into $f(x) = ax + 2$. Then, $3 = a + 2$; $a = 1$ $\therefore f(x) = x + 2$

$\qquad$ Substitute $(2, b)$ into $(x) = ax + 2$. Then, $b = 2a + 2 = 2 + 2 = 4$

$\qquad \therefore a - b = 1 - 4 = -3$

#4 Find the value of $a + b$ for which

(1) The graph of $y = ax + 2$ is translated by b along the y-axis from a graph of $y = 3x - 5$.

$\qquad y = 3x - 5 + b = ax + 2$ $\therefore a = 3$ and $-5 + b = 2$; $b = 7$

$\qquad \therefore a + b = 3 + 7 = 10$

(2) The graph is translated by a along the y-axis from a graph of $y = 2x + 4$ and passes through both point $(a + 1, -2)$ and point $\left(-\dfrac{1}{3}, b\right)$.

$\qquad$ Since the translated graph is $y = 2x + 4 + a$ and this graph passes through a point $(a + 1, -2)$,

$\qquad -2 = 2(a + 1) + 4 + a$ So, $3a = -8$; $a = -\dfrac{8}{3}$

$\qquad \therefore$ The translated graph is $y = 2x + 4 - \dfrac{8}{3} = 2x + \dfrac{4}{3}$.

$\qquad$ Since this graph passes through the point $\left(-\dfrac{1}{3}, b\right)$, $b = 2\left(-\dfrac{1}{3}\right) + \dfrac{4}{3} = \dfrac{2}{3}$

$\qquad$ Therefore, $a + b = -\dfrac{8}{3} + \dfrac{2}{3} = -2$

(3) A point $(-1, 1)$ is on the graph of $y = -2x + a$. If the graph is translated by b along the y-axis, then it will pass through the point $(3, -4)$.

$1 = -2(-1) + a$; $a = -1$ ∴ $y = -2x - 1$

The translated graph is $y = -2x - 1 + b$. So, $-4 = -2(3) - 1 + b$; $b = 3$

∴ $a + b = -1 + 3 = 2$

#5 Find the x-intercept and y-intercept.

(1) The linear function $y = ax + b$ passes through both point $(1, 2)$ and point $(-1, 4)$.

$$\begin{array}{rl} 2 = & a + b \\ +)\ 4 = & -a + b \\ \hline 6 = & 2b \ ; \ b = 3 \ \therefore a = -1 \end{array}$$

∴ $y = -x + 3$

Therefore, the x-intercept (when $y = 0$) is 3 and the y-intercept (when $x = 0$) is 3.

(2) The graph of $y = ax + b$ intersects the graph of $y = 2x + 3$ on the x-axis. It also intersects the graph of $y = -5x - 6$ on the y-axis.

the x-intercept $= -\frac{3}{2}$; $\left(-\frac{3}{2}, 0\right)$ and the y-intercept $= -6$; $(0, -6)$

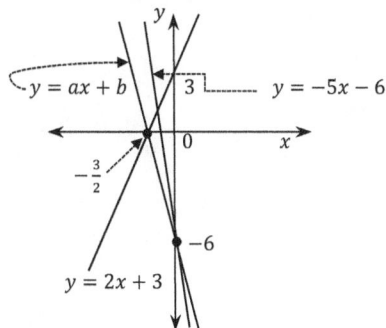

(3) The area surrounded by the graph of $y = \frac{1}{2}x + a$ $(a > 0)$, the x-axis, and the y-axis is 36.

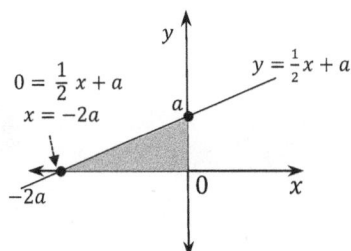

$0 = \frac{1}{2}x + a$
$x = -2a$

∴ $36 = \frac{1}{2} \cdot (2a) \cdot a = a^2$; $a = 6$ (∵ $a > 0$)

∴ $y = \frac{1}{2}x + 6$

∴ The x-intercept is -12 and the y-intercept is 6.

#6 Find the area of the polygon surrounded by

(1) A graph of $y = -\frac{2}{3}x + 2$, the x-axis, and the y-axis

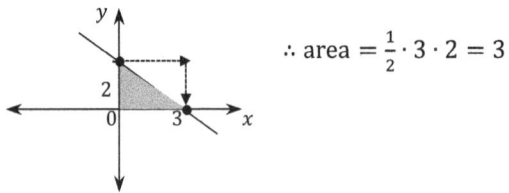

$\therefore$ area $= \frac{1}{2} \cdot 3 \cdot 2 = 3$

(2) The graphs of $y = x + 4$ and $y = -\frac{1}{2}x + 4$ and the x-axis

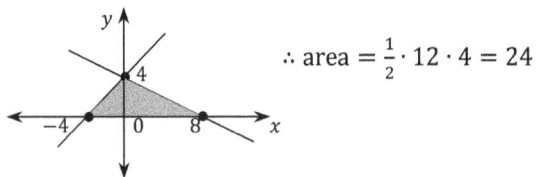

$\therefore$ area $= \frac{1}{2} \cdot 12 \cdot 4 = 24$

(3) The graphs of $y = \frac{1}{3}x + 3$ and $y = \frac{1}{9}x + 1$ and the y-axis

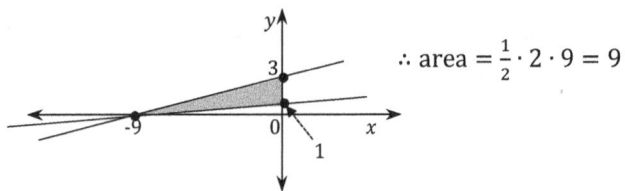

$\therefore$ area $= \frac{1}{2} \cdot 2 \cdot 9 = 9$

(4) The graphs of $x = 3$ and $y = 4$, the x-axis, and the y-axis

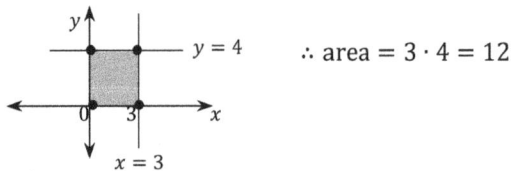

$\therefore$ area $= 3 \cdot 4 = 12$

#7 Find the slopes of the lines containing the given two points.

(1) $(2, 3)$ and $(4, -1)$

The slope $m = \frac{-1-3}{4-2} = \frac{-4}{2} = -2$

(2) $(0, 5)$ and $(-5, 0)$

The slope $m = \frac{0-5}{-5-0} = \frac{-5}{-5} = 1$

(3) $(4, -1)$ and $(2, -5)$

The slope $m = \frac{-5+1}{2-4} = \frac{-4}{-2} = 2$

(4) $(-3, 2)$ and $(0, -1)$

The slope $m = \frac{-1-2}{0-(-3)} = \frac{-3}{3} = -1$

(5) $(-2a, 0)$ and $(0, -4a), a \neq 0$

The slope $m = \frac{-4a-0}{0-(-2a)} = \frac{-4a}{2a} = -2$

#8 Find the slope m and y-intercept b of each line.

(1) $2x + 3y = 4$

$3y = -2x + 4$; $y = -\frac{2}{3}x + \frac{4}{3}$ $\quad \therefore m = -\frac{2}{3}$ and $b = \frac{4}{3}$

(2) $3y = 4x - 5$

$y = \frac{4}{3}x - \frac{5}{3}$ $\quad \therefore m = \frac{4}{3}$ and $b = -\frac{5}{3}$

(3) $x + 3y = -2$

$y = -\frac{1}{3}x - \frac{2}{3}$ $\quad \therefore m = -\frac{1}{3}$ and $b = -\frac{2}{3}$

(4) $y + 5 = 3x$

$y = 3x - 5$ $\quad \therefore m = 3$ and $b = -5$

(5) $y = 2x$

$y = 2x + 0$ $\quad \therefore m = 2$ and $b = 0$

(6) $y - 2 = 0$

$y = 2 = 0 \cdot x + 2 \quad \therefore m = 0$ and $b = 2$

#9 Find an equation in the standard form for each line.

(1) with y-intercept -3 and slope 2

$y = 2x - 3 \quad \therefore 2x - y - 3 = 0$

(2) with y-intercept 5 and slope 0

$y = 0 \cdot x + 5 \quad \therefore y - 5 = 0$

(3) with x-intercept 5 and slope $-\frac{2}{3}$

$y = -\frac{2}{3}x + b \ ; \ 0 = -\frac{2}{3} \cdot 5 + b \ ; \ b = \frac{10}{3} \ \therefore y = -\frac{2}{3}x + \frac{10}{3}$

Therefore, $2x + 3y - 10 = 0$

(4) with x-intercept -3 and slope -2

$y = -2x + b \ ; \ 0 = -2 \cdot (-3) + b \ ; \ b = -6 \ \therefore y = -2x - 6$

Therefore, $2x + y + 6 = 0$

(5) through $(1, 2)$ with slope 3

$y = 3x + b \ ; 2 = 3 \cdot 1 + b \ ; \ b = -1 \quad \therefore y = 3x - 1 \quad \therefore 3x - y - 1 = 0$

(OR $y - 2 = 3(x - 1) \ ; \ y = 3x - 1 \quad \therefore 3x - y - 1 = 0$)

(6) through $(3, -4)$ with slope -2

$y = -2x + b \ ; -4 = -2 \cdot 3 + b \ ; \ b = 2 \quad \therefore y = -2x + 2 \quad \therefore 2x + y - 2 = 0$

(OR $y + 4 = -2(x - 3) \ ; \ y = -2x + 2 \quad \therefore 2x + y - 2 = 0$)

(7) through $(2, 3)$ with undefined slope

no slope $\Rightarrow$ no change in $x \ \therefore x = 2 \ ; \ x - 2 = 0$

(8) through $(-2, 3)$ with y-intercept -1

$y = \frac{-4}{2}x - 1 = -2x - 1 \quad \therefore 2x + y + 1 = 0$

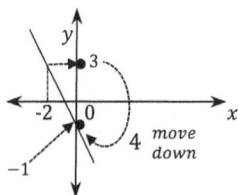

(9) through $(2, 4)$ with x-intercept -5

$$y = \frac{4}{7}x + b \; ; \; 0 = \frac{4}{7} \cdot (-5) + b; \; b = \frac{20}{7} \quad \therefore y = \frac{4}{7}x + \frac{20}{7} \quad \therefore 4x - 7y + 20 = 0$$

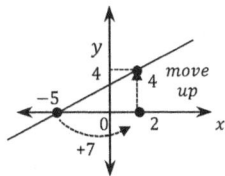

(10) through $(3, 1)$ and $(-2, 4)$

$$m = \frac{4-1}{-2-3} = -\frac{3}{5} \quad \therefore \; y = -\frac{3}{5}x + b$$

So, $\quad 1 = -\frac{3}{5} \cdot 3 + b \; ; \; b = \frac{14}{5} \quad \therefore \; y = -\frac{3}{5}x + \frac{14}{5}$

Therefore, $\quad 3x + 5y - 14 = 0$

(OR Using Point-Slope form, $y - 1 = -\frac{3}{5}(x - 3) \; ; \; 5y - 5 = -3x + 9 \quad \therefore \; 3x + 5y - 14 = 0$)

(11) through $(-2, -3)$ and $(-1, 5)$

$$m = \frac{5+3}{-1+2} = 8 \quad \therefore \; y = 8x + b$$

So, $\quad 5 = 8 \cdot (-1) + b \; ; \; b = 13 \quad \therefore \; y = 8x + 13$

Therefore, $\quad 8x - y + 13 = 0$

(OR Using Point-Slope form, $y + 3 = 8(x + 2) \quad \therefore \; 8x - y + 13 = 0$)

(12) with x-intercept -3 and y-intercept 3

$$m = \frac{0-3}{-3-0} = 1 \quad \therefore \; y = x + 3 \quad \therefore x - y + 3 = 0$$

(13) with x-intercept $\frac{3}{2}$ and y-intercept -4

$$m = \frac{-4-0}{0-\frac{3}{2}} = \frac{-4}{-\frac{3}{2}} = \frac{8}{3} \quad \therefore \; y = \frac{8}{3}x - 4 \quad \therefore 8x - 3y - 12 = 0$$

(14) Vertical line through $(-1, 2)$

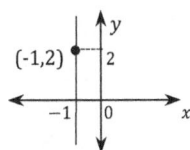

$$x = -1 \quad \therefore x + 1 = 0$$

(15) Horizontal line through $(3, -4)$

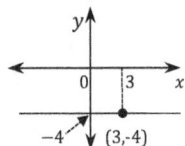

$$y = -4 \quad \therefore y + 4 = 0$$

#10 Find an equation for the line through $(2, 3)$ which is

(1) parallel to the line $y = 2x - 5$

$y = 2x + b$; $3 = 2 \cdot 2 + b$; $b = -1$ ∴ $y = 2x - 1$ ∴ $2x - y - 1 = 0$

(2) parallel to the line $y = -3x + 1$

$y = -3x + b$; $3 = -3 \cdot 2 + b$; $b = 9$ ∴ $y = -3x + 9$ ∴ $3x + y - 9 = 0$

(3) parallel to the line $x = 4$

$x = 2$ ∴ $x - 2 = 0$

(4) parallel to the line $y = -2$

$y = 3$ ∴ $y - 3 = 0$

(5) parallel to the line $3x + 4y = 5$

$y = -\frac{3}{4}x + \frac{5}{4}$ ∴ $y = -\frac{3}{4}x + b$

$3 = -\frac{3}{4} \cdot 2 + b$; $b = \frac{9}{2}$ ∴ $y = -\frac{3}{4}x + \frac{9}{2}$

Therefore, $3x + 4y - 18 = 0$

(6) perpendicular to the line $y = \frac{2}{3}x - 1$

$y = -\frac{3}{2}x + b$

$3 = -\frac{3}{2} \cdot 2 + b$; $b = 6$ ∴ $y = -\frac{3}{2}x + 6$

Therefore, $3x + 2y - 12 = 0$

(7) perpendicular to the line $x + 3y = -3$

$x + 3y = -3$ ⟹ $y = -\frac{1}{3}x - 1$

∴ $y = 3x + b$

$3 = 3 \cdot 2 + b$; $b = -3$ ∴ $y = 3x - 3$

Therefore, $3x - y - 3 = 0$

(8) perpendicular to the line $x = 5$

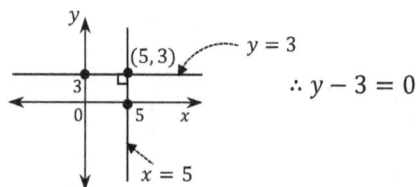

∴ $y - 3 = 0$

(9) perpendicular to the line $y = -2$

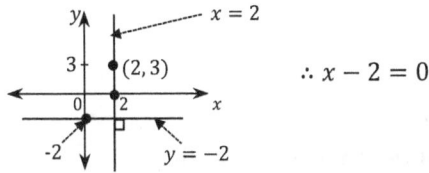

$\therefore x - 2 = 0$

#11 Find the value of a for the following lines

(1) through $(2, 3)$ **and** $(1, -a)$ **with slope 2**

$2 = \frac{-a-3}{1-2} = \frac{-a-3}{-1} = a + 3 \quad \therefore a = -1$

(2) through $(2a - 1, -2)$ **and** $(-1, 1)$ **with slope** -2

$-2 = \frac{1+2}{-1-2a+1} = \frac{3}{-2a} \; ; \; 4a = 3 \quad \therefore a = \frac{3}{4}$

(3) through $(1, -2)$, $(-3, 2)$, **and** $(-a + 1, -5)$

$m = \frac{2+2}{-3-1} = \frac{4}{-4} = -1 \quad \therefore y = -x + b$

So, $-2 = -1 + b \; ; \; b = -1 \quad \therefore y = -x - 1$

$-5 = a - 1 - 1 = a - 2 \quad \therefore a = -3$

(4) through $(2a + 1, -4)$, $(2, 5)$, **and** $(2, -3)$

$(2, 5)$ and $(2, -3) \Rightarrow$ no change in x ; no slope ; vertical line

$\therefore 2a + 1 = 2 \quad \therefore a = \frac{1}{2}$

(5) through $(-3, 3)$, $(3, a - 1)$, **and** $(0, 3)$

$(-3, 3)$ and $(0, 3) \Rightarrow$ no change in y ; slope is 0 ; horizontal line

$\therefore a - 1 = 3 \quad \therefore a = 4$

(6) through $(a, 2a - 3)$ **and** $(-a - 1, 3 + 4a)$ **and parallel to the** x**-axis**

parallel to x-axis $\Rightarrow$ horizontal line

$\therefore 2a - 3 = 3 + 4a \; ; \; 2a = -6 \quad \therefore a = -3$

(7) through $(-3a + 1, -5)$ **and** $(2a - 1, a + 3)$ **and perpendicular to the** x**-axis**

perpendicular to x-axis $\Rightarrow$ vertical line

$\therefore -3a + 1 = 2a - 1 \; ; \; 5a = 2 \quad \therefore a = \frac{2}{5}$,

(8) through $(3, -2a)$ and $(2a - 1, -3a + 2)$ and parallel to the y-axis

parallel to y-axis $\Rightarrow$ vertical line

$\therefore 3 = 2a - 1$; $2a = 4$ $\therefore a = 2$

(9) through $(-1, 5)$ and $(2, -4)$ and parallel to the line $ax + 3y + 5 = 0$

$m = \frac{-4-5}{2+1} = \frac{-9}{3} = -3$ $\therefore y = -3x + b$

$ax + 3y + 5 = 0 \Rightarrow 3y = -ax - 5 \Rightarrow y = -\frac{a}{3}x - \frac{5}{3}$

$\therefore -\frac{a}{3} = -3$ $\therefore a = 9$

#12 Find the value of a such that the line $ax + 2y = 5$

(1) is parallel to the line $2x + 3y = -2$.

the same slope

$3y = -2x - 2$; $y = -\frac{2}{3}x - \frac{2}{3}$ $\therefore$ slope $= -\frac{2}{3}$

$ax + 2y = 5 \Rightarrow y = -\frac{1}{2}ax + \frac{5}{2}$ $\therefore$ slope $= -\frac{1}{2}a$

Therefore, $-\frac{2}{3} = -\frac{1}{2}a$ $\therefore a = \frac{4}{3}$

(2) is perpendicular to the line $y = -2x + 3$.

negative reciprocals of each other

Since $-2 \cdot \left(-\frac{1}{2}a\right) = -1$, $a = -1$

(3) coincides with the line $6y = -4x + 15$.

$6y = -4x + 15 \Rightarrow 4x + 6y = 15 \Rightarrow \frac{4}{3}x + 2y = 5$

$\therefore a = \frac{4}{3}$

#13 Find the value of ab for which

(1) the system $\begin{cases} x - 3y = a \\ 2x + by = 3 \end{cases}$ has the intersection point $(2, 3)$.

Substitute $(2, 3)$ into each equation.

Then, $2 - 9 = a$; $a = -7$ and $4 + 3b = 3$; $b = -\frac{1}{3}$

$\therefore ab = \frac{7}{3}$

(2) the system $\begin{cases} -ax + by = 4 \\ 2ax + 3by = 2 \end{cases}$ **has the intersection point** $(-1, 2)$.

Substitute $(-1, 2)$ into each equation.

Then, $\begin{cases} a + 2b = 4 \\ -2a + 6b = 2 \end{cases}$

$2a + 4b = 8$

$+) \underline{-2a + 6b = 2}$

$\qquad\qquad 10b = 10 \ ; b = 1$

$\therefore a = 4 - 2b = 4 - 2 = 2$

$\therefore ab = 2$

(3) the system $\begin{cases} px + y = 3 \\ 2x - 3y = q \end{cases}$ **has no intersection when** $p = a, \ q \neq b$.

parallel $\quad \therefore \dfrac{p}{2} = -\dfrac{1}{3} \neq \dfrac{3}{q} \quad \therefore p = -\dfrac{2}{3}$ and $q \neq -9$

$\therefore a = -\dfrac{2}{3}, b = -9 \quad \therefore ab = 6$

(4) the system $\begin{cases} 2ax + 4y = -3 \\ 3x + 6y = 2b \end{cases}$ **has unlimited numbers of intersections.**

$\dfrac{2a}{3} = \dfrac{4}{6} = \dfrac{-3}{2b} \quad \therefore 2a = 2 \ ; a = 1$ and $4b = -9 \ ; \ b = -\dfrac{9}{4}$

$\therefore ab = -\dfrac{9}{4}$

#14 Find the value of a such that :

(1) the system $\begin{cases} ax + y = -2 \\ -3x + 2y = 4 \end{cases}$ **has no solution.**

parallel

$-\dfrac{a}{3} = \dfrac{1}{2} \neq -\dfrac{2}{4} \quad \therefore a = -\dfrac{3}{2}$

(2) the system $\begin{cases} 2x - ay + 3 = 0 \\ x + 3y - 2 = 0 \\ 2x + y + 1 = 0 \end{cases}$ **has one solution.**

$\begin{cases} 2x - ay + 3 = 0 \cdots\cdots ① \\ x + 3y - 2 = 0 \cdots\cdots ② \\ 2x + y + 1 = 0 \cdots\cdots ③ \end{cases}$

$\Rightarrow ① - ③ \ ; (-a - 1)y + 2 = 0$

$\qquad 2 \cdot ② - ③ \ ; 5y - 5 = 0 \ ; y = 1$

$\therefore (-a - 1) + 2 = 0 \quad \therefore a = 1$

(3) the system $\begin{cases} x - 3y = 2 \\ 2x + y = -3 \end{cases}$ has a solution $(2a, -1)$.

$$2x - 6y = 4$$
$$-)\ \underline{2x + y = -3}$$
$$-7y = 7 \ ; y = -1$$
$$\therefore x = 3y + 2 = -3 + 2 = -1 \quad \therefore 2a = -1 \quad \therefore a = -\frac{1}{2}$$

(4) the line $2ax + 3y - 1 = 0$ passes through the intersection of the system $\begin{cases} x - 2y = 3 \\ 2x + 2y = 1 \end{cases}$.

$$2x - 4y = 6 \qquad\qquad x - 2y = 3$$
$$-)\ \underline{2x + 2y = 1} \qquad +)\ \underline{2x + 2y = 1}$$
$$-6y = 5 \ ; \ y = -\frac{5}{6} \qquad 3x = 4 \ ; \ x = \frac{4}{3}$$
$$\therefore 2a \cdot \frac{4}{3} + 3 \cdot \left(-\frac{5}{6}\right) - 1 = 0 \quad ; \quad \frac{8}{3}a = \frac{7}{2} \quad \therefore a = \frac{21}{16}$$

#15 Find the equation of each line such that

(1) the line passes through the intersection of the system $\begin{cases} x + 2y = 3 \\ 3x + y = -2 \end{cases}$

and runs parallel to the y-axis.

$$x + 2y = 3$$
$$-)\ \underline{6x + 2y = -4}$$
$$-5x = 7 \ ; \ x = -\frac{7}{5} \quad \therefore 2y = -x + 3 = \frac{7}{5} + 3 = \frac{22}{5} \quad \therefore y = \frac{11}{5}$$
$$\therefore \text{ The intersection is } \left(-\frac{7}{5}, \frac{11}{5}\right).$$

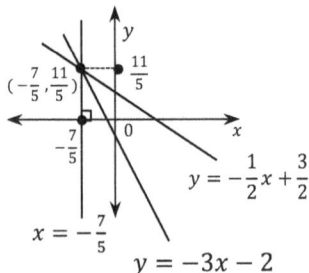

parallel to the y-axis
$$\Rightarrow x = -\frac{7}{5} \quad \therefore 5x + 7 = 0$$

(2) the line passes through the intersection of the system $\begin{cases} -x + y + 2 = 0 \\ 2x + y - 3 = 0 \end{cases}$

and runs perpendicular to the x-axis.

$$-x + y + 2 = 0$$
$$-) \ \underline{2x + y - 3 = 0}$$
$$-3x \quad + 5 = 0 \ ; \ x = \frac{5}{3} \quad \therefore \ y = x - 2 = \frac{5}{3} - 2 = -\frac{1}{3}$$

$\therefore$ The intersection is $\left(\frac{5}{3}, -\frac{1}{3} \right)$.

$\therefore$ Perpendicular to the x-axis $\Rightarrow x = \frac{5}{3}$

$\therefore 3x - 5 = 0$

(3) the line passes through the intersection of the system $\begin{cases} 2x - y + 3 = 0 \\ x + 2y + 4 = 0 \end{cases}$

and runs parallel to the line $3x + 2y = 5$.

$$2x - y + 3 = 0$$
$$-) \ \underline{2x + 4y + 8 = 0}$$
$$-5y - 5 = 0 \ ; \ y = -1$$

$\therefore \ x = -2y - 4 = -2$

$\therefore$ The intersection is $(-2, -1)$.

$3x + 2y = 5 \Rightarrow y = -\frac{3}{2}x + \frac{5}{2} \quad \therefore m = -\frac{3}{2}$

$\therefore \ y = -\frac{3}{2}x + b \ ; \ -1 = -\frac{3}{2} \cdot (-2) + b \ ; \ b = -4$

$\therefore \ y = -\frac{3}{2}x - 4$

$\therefore 3x + 2y + 8 = 0$

#16 Find the area of the polygon surrounded by two lines ($3x + 4y - 16 = 0$ and $3x - 2y - 10 = 0$), x-axis, and y-axis.

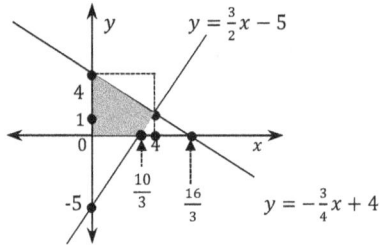

$$\begin{cases} 3x + 4y - 16 = 0 \\ 3x - 2y - 10 = 0 \end{cases} \Rightarrow \begin{cases} 3x + 4y - 16 = 0 \\ 6x - 4y - 20 = 0 \end{cases} \Rightarrow 9x - 36 = 0 \Rightarrow x = 4 \quad \therefore \text{The intersection point is } (4, 1).$$

So, the area is $\frac{1}{2} \cdot 9 \cdot 4 - \frac{1}{2} \cdot 5 \cdot \frac{10}{3} = 18 - \frac{25}{3} = \frac{54-25}{3} = \frac{29}{3}$.

#17 The area of a polygon surrounded by $y = x$, $y = ax + b$ ($b > 0$) which has x-intercept 6, and the x-axis is 12. Find the area of a polygon surrounded by those two lines and the y-axis.

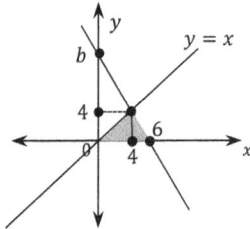

$\because 12 = \frac{1}{2} \cdot 6 \cdot \text{height} \quad ; \quad \text{height} = 4 \quad \therefore (4, 4) \text{ is on } y = x$

Passes through $(4, 4)$ and $(6, 0) \quad \Rightarrow \quad$ The slope is $= \frac{0-4}{6-4} = -2$

$\therefore \quad y = -2x + b \; ; \; 0 = -2 \cdot 6 + b \; ; \; b = 12 \quad \therefore \quad y = -2x + 12$

Therefore, the area is $\frac{1}{2} \cdot 12 \cdot 4 = 24$.

#18 The perimeter of a rectangle with the length 5 inches and the width x inches is y square inches. Find the relationship between x and y.

$y = 2 \cdot 5 + 2 \cdot x \qquad \therefore y = 2x + 10$

#19 Richard drives a car 15 miles total, from place A to place B. He begins at a speed of 30 miles per hour. x minutes after departing, he has y miles more to go to arrive at place B. Find the relationship between x and y.

Since the distance for x minutes is $30 \cdot \frac{x}{60}$, $30 \cdot \frac{x}{60} + y = 15$.

So, $y = 15 - 30 \cdot \frac{x}{60} = 15 - \frac{x}{2}$ $\therefore y = -\frac{1}{2}x + 15$

#20 Richard and Nichole drive toward each other from opposite starting points 4 miles apart. Richard drives at a speed of 40 miles per hour and Nichole drives at a speed of 35 miles per hour. After x minutes, the distance between the two is y miles. Find the relationship between x and y.

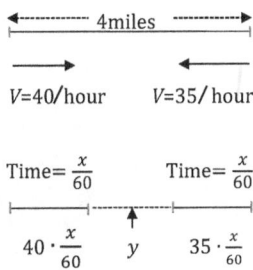

$$y = 4 - \left(40 \cdot \frac{x}{60} + 35 \cdot \frac{x}{60}\right)$$

$$= 4 - \frac{15x}{12} = 4 - \frac{5}{4}x$$

$$\therefore y = -\frac{5}{4}x + 4 \quad \left(0 \leq x \leq 3\frac{1}{5}\right)$$

$$\left(\because \text{Since } y \geq 0, -\frac{5}{4}x + 4 \geq 0 \; ; \; \frac{5}{4}x \leq 4 \; ; \; x \leq \frac{16}{5} = \left(3\frac{1}{5}\right)\right)$$

#21 Graph the following lines

(1) $y = -|x|$

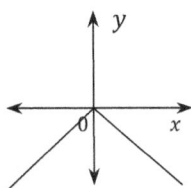

$$y = -|x| = \begin{cases} x \geq 0 \Rightarrow y = -x \\ x < 0 \Rightarrow y = x \end{cases}$$

Domain = all real numbers

Range = $\{ y|\, y \leq 0 \}$: all non-positive real numbers

(2) $y = -|x + 3| + 2$

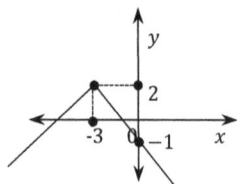

$$y = -|x + 3| + 2 = \begin{cases} x + 3 \geq 0 & \Rightarrow y = -x - 1 \\ x + 3 < 0 & \Rightarrow y = x + 5 \end{cases}$$

Domain = all real numbers

Range = $\{ y|\, y \leq 2 \}$

(3) $y = |x| + x$

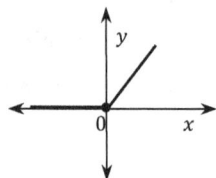

$$y = |x| + x = \begin{cases} x \geq 0 & \Rightarrow y = 2x \\ x < 0 & \Rightarrow y = 0 \end{cases}$$

Domain = all real numbers

Range = $\{ y|\, y \geq 0 \}$

(4) $|y - 1| = x + 2$

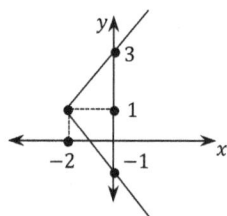

$$\begin{cases} y - 1 \geq 0 & \Rightarrow y - 1 = x + 2 \Rightarrow y = x + 3 \\ y - 1 < 0 & \Rightarrow -(y - 1) = x + 2 \Rightarrow y = -x - 1 \end{cases}$$

Domain = $\{ x|\, x \geq -2 \}$

Range = all real numbers

Solutions for Chapter 3

#1 Identify the quadratic functions by marking 0 or $\times$.

(1) $y = \frac{1}{2}x^2 + 1$; 0

(2) $y = 2x^2 - (3 + 2x^2)$; $\times$

(3) $y = \frac{1}{x^2} + 1$; $\times$

(4) $y = x^2 - (x+1)^2$; $\times$

(5) $y = x(x+1)$; 0

(6) $y = 2x^2 - x^2 + 1$; 0

(7) $y = \frac{(x+1)^2}{3}$; 0

(8) $2x^2 + 3x + 1$; $\times$

(9) $y = 2$; $\times$

(10) $y = 3x + 1$; $\times$

#2 Find the following values of the quadratic function $f(x) = x^2 - 2x - 1$.

(1) $f(0) = -1$

(2) $f(-1) = 1 + 2 - 1 = 2$

(3) $f(2) + f(-2) = 4 - 4 - 1 + 4 + 4 - 1 = 6$

(4) $f\left(-\frac{1}{2}\right) = \frac{1}{4} + 1 - 1 = \frac{1}{4}$

(5) $2f(1) = 2(1 - 2 - 1) = -4$

#3 Find the value of $a + b$ for the quadratic equation $f(x) = -\frac{1}{2}x^2 + a$.

(1) $f(1) = -3$ and $f(-2) = b$

Since $f(1) = -\frac{1}{2} + a = -3$, $a = -3 + \frac{1}{2} = -\frac{5}{2}$

Since $f(-2) = -\frac{1}{2} \cdot 4 + a = b$, $b = -2 - \frac{5}{2} = -\frac{9}{2}$

$\therefore a + b = -\frac{5}{2} - \frac{9}{2} = -7$

(2) $f(-1) = 1$ and $\frac{1}{2}f(0) = 2b$

Since $f(-1) = -\frac{1}{2} + a = 1$, $a = 1 + \frac{1}{2} = \frac{3}{2}$

Since $f(0) = a$, $\frac{1}{2}f(0) = \frac{1}{2}a = 2b$; $b = \frac{3}{8}$

$\therefore a + b = \frac{3}{2} + \frac{3}{8} = \frac{15}{8}$

(3) $\frac{f(1)+f(-1)}{2} = -\frac{1}{4}$ and $f(2) = -b$

Since $f(1) = -\frac{1}{2} + a$ and $f(-1) = -\frac{1}{2} + a$, $f(1) + f(-1) = -1 + 2a$

So, $\frac{-1+2a}{2} = -\frac{1}{4}$; $-1 + 2a = -\frac{1}{2}$; $a = \frac{1}{4}$

Since $f(2) = -2 + a = -b$, $b = 2 - a = 2 - \frac{1}{4} = \frac{7}{4}$

$\therefore a + b = \frac{1}{4} + \frac{7}{4} = 2$

#4 Find the vertex and the axis of symmetry for the following parabolas

(1) $y = 2x^2 - 4x = 2(x^2 - 2x) = 2((x-1)^2 - 1) = 2(x-1)^2 - 2$

$\therefore$ vertex : $(1, -2)$ and axis of symmetry : $x = 1$

(2) $y = x^2 - 2x - 3 = (x-1)^2 - 1 - 3 = (x-1)^2 - 4$

$\therefore$ vertex : $(1, -4)$ and axis of symmetry : $x = 1$

(3) $y = -x^2 - 2x + 2 = -(x^2 + 2x - 2) = -(x+1)^2 + 3$

$\therefore$ vertex : $(-1, 3)$ and axis of symmetry : $x = -1$

(4) $y = -\frac{1}{2}x^2 + 1 = -\frac{1}{2}(x-0)^2 + 1$

$\therefore$ vertex : $(0, 1)$ and axis of symmetry : $x = 0$

(5) $y - 3 = 2(x-2)^2$; $y = 2(x-2)^2 + 3$

$\therefore$ vertex : $(2, 3)$ and axis of symmetry : $x = 2$

#5 Identify the equations of the functions whose graphs are translated from the graph of $y = \frac{1}{2}x^2$ in the following ways

    (1) Translated -2 units along the x-axis

       $y = \frac{1}{2}(x+2)^2$

    (2) Translated 2 units along the y-axis

       $y - 2 = \frac{1}{2}x^2$; $y = \frac{1}{2}x^2 + 2$

    (3) Translated 1 unit along the x-axis and -1 unit along the y-axis

       $y + 1 = \frac{1}{2}(x-1)^2$; $y = \frac{1}{2}(x-1)^2 - 1$

    (4) Translated -3 units along the x-axis and -4 units along the y-axis

       $y + 4 = \frac{1}{2}(x+3)^2$; $y = \frac{1}{2}(x+3)^2 - 4$

    (5) Translated m units along the x-axis and n units along the y-axis

       $y - n = \frac{1}{2}(x-m)^2$; $y = \frac{1}{2}(x-m)^2 + n$

#6 Find the value of a for which

    (1) The graph of $y = ax^2$ passes through one point $(2, -2)$.

       $-2 = 4a$; $a = -\frac{1}{2}$

    (2) The graph of $y = \left(x - \frac{1}{2}\right)^2 + a$ passes through one point $(-1, 3)$.

       $3 = (-1 - \frac{1}{2})^2 + a$; $3 = \frac{9}{4} + a$; $a = \frac{3}{4}$

    (3) The graph of $y = \left(x + \frac{a}{2}\right)^2 + 3$ has the vertex $(-5, 3)$.

       $\left(-\frac{a}{2}, 3\right) = (-5, 3)$; $\frac{a}{2} = 5$; $a = 10$

    (4) The graph of $y = \left(x - \frac{2a}{3}\right)^2 - 2$ has been translated from the graph of $y = x^2 - 2$,

       -4 units along the x-axis .

       $y = (x+4)^2 - 2 = \left(x - \frac{2a}{3}\right)^2 - 2$; $\frac{2a}{3} = -4$; $a = -6$

(5) The graph of $y = 2(x + 3a - 1)^2 + 1$ **has the y-axis as the axis of symmetry.**

axis of symmetry : $x = -3a + 1$

y-axis : $x = 0$

So, $-3a + 1 = 0$; $a = \dfrac{1}{3}$

#7 Find the value of ab when the parabola $y = -ax^2 + b$

(1) Passes through $(-1, 2)$ and $(3, -2)$

$$2 = -a + b$$
$$-) \underline{-2 = -9a + b}$$
$$4 = 8a \;;\; a = \dfrac{1}{2} \qquad \therefore b = \dfrac{1}{2} + 2 = \dfrac{5}{2}$$

So, $ab = \dfrac{5}{4}$

(2) Passes through $(1, 2)$ and $(-2, -4)$

$$2 = -a + b$$
$$-) \underline{-4 = -4a + b}$$
$$6 = 3a \;;\; a = 2 \qquad \therefore b = 2 + 2 = 4$$

So, $ab = 8$

#8 Find the value of $a + b$ for the following graphs of quadratic functions

(1) $y = 2x^2 - x + 3$ passes through the two points $(1, a)$ and $(-2, -b)$.

$a = 2 - 1 + 3 = 4$ and $-b = 8 + 2 + 3 = 13$ ($; b = -13$)

$\therefore a + b = 4 - 13 = -9$

(2) $y = -ax^2 + 2x - 1$ passes through the two points $(1, b)$ and $(-1, a)$.

$b = -a + 2 - 1 = -a + 1$ and $a = -a - 2 - 1 = -a - 3$ ($; a = -\dfrac{3}{2}$)

$\therefore a + b = -\dfrac{3}{2} + (\dfrac{3}{2} + 1) = 1$

(3) $y = -x^2 + 2ax + 3$ passes through the two points $(-1, 0)$ and $(2, b)$.

$0 = -1 - 2a + 3$; $2a = 2$; $a = 1$

$b = -4 + 4a + 3 = 4a - 1 = 3$

$\therefore a + b = 4$

#9 For any constants m, n, the parabola $y = \frac{1}{2}(x - m)^2 + n$ is translated from $y = \frac{1}{2}x^2$. Give the conditions for m and n for the following parabola

(1)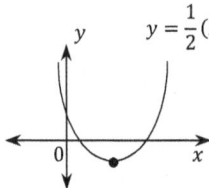

$m > 0, \ n < 0$

(2)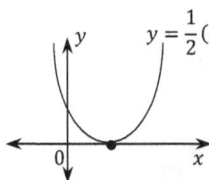

$m > 0, \ n = 0$

(3)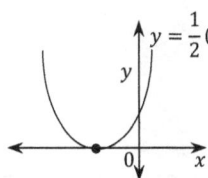

$m < 0, \ n = 0$

(4)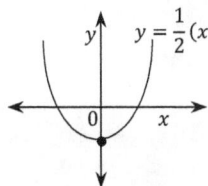

$m = 0, \ n < 0$

(5)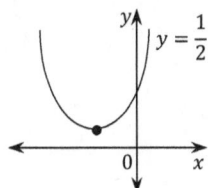

$m < 0, \ n > 0$

(6)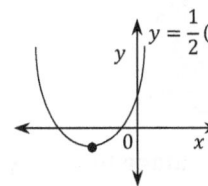

$m < 0, \ n < 0$

(7)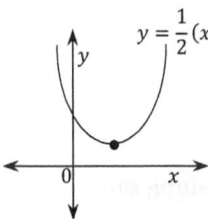

$m > 0, \ n > 0$

#10 Find an equation for the resulting quadratic function when

(1) The parabola $y = 2x^2 - 6x + 5$ is translated 2 units along the x-axis and -1 unit along the y-axis.

$$y = 2x^2 - 6x + 5 = 2(x^2 - 3x) + 5 = 2((x - \tfrac{3}{2})^2 - \tfrac{9}{4}) + 5$$

$$= 2\left(x - \tfrac{3}{2}\right)^2 - \tfrac{9}{2} + 5 = 2\left(x - \tfrac{3}{2}\right)^2 + \tfrac{1}{2}$$

Substitute $x - 2$ into x and $y + 1$ into y

$$\Rightarrow y + 1 = 2\left(x - 2 - \tfrac{3}{2}\right)^2 + \tfrac{1}{2} \quad \therefore \ y = 2\left(x - \tfrac{7}{2}\right)^2 - \tfrac{1}{2}$$

(OR vertex : $\left(\tfrac{3}{2}, \tfrac{1}{2}\right)$ $\therefore$ new vertex : $\left(\tfrac{3}{2} + 2, \tfrac{1}{2} - 1\right) = \left(\tfrac{7}{2}, -\tfrac{1}{2}\right)$

Therefore, $y = 2\left(x - \tfrac{7}{2}\right)^2 - \tfrac{1}{2}$)

(2) The parabola $y = -3x^2 + 2x - 2$ is translated -2 units along the x-axis and 3 units along the y-axis.

$$y = -3x^2 + 2x - 2 = -3(x^2 - \tfrac{2}{3}x) - 2 = -3((x - \tfrac{1}{3})^2 - \tfrac{1}{9}) - 2$$

$$= -3\left(x - \tfrac{1}{3}\right)^2 + \tfrac{1}{3} - 2 = -3\left(x - \tfrac{1}{3}\right)^2 - \tfrac{5}{3}$$

Substitute $x + 2$ into x and $y - 3$ into y

$$\Rightarrow y - 3 = -3\left(x + 2 - \tfrac{1}{3}\right)^2 - \tfrac{5}{3} \quad \therefore \ y = -3\left(x + \tfrac{5}{3}\right)^2 + \tfrac{4}{3}$$

(OR vertex : $\left(\tfrac{1}{3}, -\tfrac{5}{3}\right)$ $\therefore$ new vertex : $\left(\tfrac{1}{3} - 2, -\tfrac{5}{3} + 3\right) = \left(-\tfrac{5}{3}, \tfrac{4}{3}\right)$

Therefore, $y = -3\left(x + \tfrac{5}{3}\right)^2 + \tfrac{4}{3}$)

(3) The parabola $y = -\tfrac{1}{2}(x + 2)^2 - 1$ is a symmetrical transformation along the x-axis.

$$-y = -\tfrac{1}{2}(x + 2)^2 - 1 \ ; \ \ ; y = \tfrac{1}{2}(x + 2)^2 + 1$$

(4) The parabola $y = \tfrac{1}{2}(x + 2)^2 + 1$ is a symmetrical transformation along the y-axis.

$$y = \tfrac{1}{2}(-x + 2)^2 + 1 \ ; \ y = \tfrac{1}{2}(x - 2)^2 + 1$$

#11 Find the equation of the parabolas A and B on the graph. Both are transformations of the parabola $y = \frac{1}{2}x^2$.

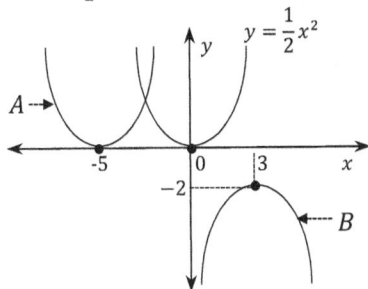

Since A is translated -5 units along the x-axis, $A : y = \frac{1}{2}(x+5)^2$

Since B is a symmetrical transformation along the x-axis and then translated 3 units along the x-axis -2 units along the y-axis,

$$-y = \frac{1}{2}x^2 \ ; \ y = -\frac{1}{2}x^2 \text{ (symmetry)} \ ; \ y + 2 = -\frac{1}{2}(x-3)^2 \text{ (transformation)}$$

$$\therefore \ B : y = -\frac{1}{2}(x-3)^2 - 2$$

#12 State how the following parabolas have been translated from $y = -x^2 + 3x - 2$.

(1) $y = -x^2 + 4x + 3$

$$y = -x^2 + 4x + 3 = -(x^2 - 4x) + 3 = -(x-2)^2 + 4 + 3 = -(x-2)^2 + 7$$

Since $y = -x^2 + 3x - 2 = -(x^2 - 3x) - 2 = -\left(x - \frac{3}{2}\right)^2 + \frac{9}{4} - 2 = -\left(x - \frac{3}{2}\right)^2 + \frac{1}{4}$,

vertex : $\left(\frac{3}{2}, \frac{1}{4}\right)$

Since $\frac{3}{2} + m = 2$ and $\frac{1}{4} + n = 7$, $m = \frac{1}{2}$ and $n = \frac{27}{4}$

$\therefore \ \frac{1}{2}$ unit along the x-axis and $\frac{27}{4}$ units along the y-axis.

(2) $y = -x^2 + \frac{1}{2}x - 1$

$$y = -x^2 + \frac{1}{2}x - 1 = -\left(x^2 - \frac{1}{2}x\right) - 1 = -\left(x - \frac{1}{4}\right)^2 + \frac{1}{16} - 1 = -\left(x - \frac{1}{4}\right)^2 - \frac{15}{16}$$

Since $\frac{3}{2} + m = \frac{1}{4}$ and $\frac{1}{4} + n = -\frac{15}{16}$, $m = -\frac{5}{4}$ and $n = -\frac{19}{16}$

$\therefore \ -\frac{5}{4}$ units along the x-axis and $-\frac{19}{16}$ units along the y-axis.

(3) $y = -x^2 + 4x$

$y = -x^2 + 4x = -(x-2)^2 + 4$; vertex : $(2, 4)$

Since $\frac{3}{2} + m = 2$ and $\frac{1}{4} + n = 4$, $m = \frac{1}{2}$ and $n = \frac{15}{4}$

$\therefore$ $\frac{1}{2}$ unit along the x-axis and $\frac{15}{4}$ units along the y-axis.

(4) $y = -x^2 + 3x + 2$

$y = -x^2 + 3x + 2 = -(x^2 - 3x) + 2 = -\left(x - \frac{3}{2}\right)^2 + \frac{9}{4} + 2 = -\left(x - \frac{3}{2}\right)^2 + \frac{17}{4}$

Since $\frac{3}{2} + m = \frac{3}{2}$ and $\frac{1}{4} + n = \frac{17}{4}$, $m = 0$ and $n = 4$

$\therefore$ 4 units along the y-axis.

#13 Find the vertex, axis of symmetry, and intercepts for the following quadratic functions

(1) $y = 2x^2 + 3x + 1$

$y = 2x^2 + 3x + 1 = 2\left(x^2 + \frac{3}{2}x\right) + 1 = 2\left(x + \frac{3}{4}\right)^2 - \frac{9}{16} \cdot 2 + 1 = 2\left(x + \frac{3}{4}\right)^2 - \frac{1}{8}$

$\therefore$ vertex : $\left(-\frac{3}{4}, -\frac{1}{8}\right)$ and axis of symmetry : $x = -\frac{3}{4}$

If $y = 0$, then $0 = 2\left(x + \frac{3}{4}\right)^2 - \frac{1}{8}$; $\left(x + \frac{3}{4}\right)^2 = \frac{1}{16}$; $x = -\frac{3}{4} \pm \frac{1}{4}$ $\therefore x = -\frac{1}{2}$ or $x = -1$

$\therefore$ The x-intercepts are $x = -\frac{1}{2}$ and $x = -1$

(OR $2x^2 + 3x + 1 = 0$; $x = \frac{-3 \pm \sqrt{3^2 - 4 \cdot 2 \cdot 1}}{2 \cdot 2} = \frac{-3 \pm 1}{4}$ $\therefore x = -\frac{1}{2}$ or $x = -1$)

If $x = 0$, then $y = 1$

$\therefore$ The y-intercept : $y = 1$

(2) $y = -x^2 + 2x + 3$

$y = -x^2 + 2x + 3 = -(x^2 - 2x) + 3 = -(x - 1)^2 + 1 + 3 = -(x - 1)^2 + 4$

$\therefore$ vertex : $(1, 4)$ and axis of symmetry : $x = 1$

If $y = 0$, then $0 = -(x - 1)^2 + 4$; $(x - 1)^2 = 4$; $x = 1 \pm 2$ $\therefore x = 3$ or $x = -1$

$\therefore$ The x-intercepts are $x = 3$ and $x = -1$

(OR $-x^2 + 2x + 3 = 0$; $x^2 - 2x - 3 = 0$; $(x - 3)(x + 1) = 0$ $\therefore x = 3$ or $x = -1$)

If $x = 0$, then $y = 3$ $\therefore$ The y-intercept : $y = 3$

(3) $y = -3x^2 - 3x$

$$y = -3x^2 - 3x = -3(x^2 + x) = -3\left(x + \frac{1}{2}\right)^2 + \frac{3}{4}$$

$\therefore$ vertex: $\left(-\frac{1}{2}, \frac{3}{4}\right)$ and axis of symmetry: $x = -\frac{1}{2}$

If $y = 0$, then $0 = -3\left(x + \frac{1}{2}\right)^2 + \frac{3}{4}$; $3\left(x + \frac{1}{2}\right)^2 = \frac{3}{4}$; $\left(x + \frac{1}{2}\right)^2 = \frac{1}{4}$; $x = -\frac{1}{2} \pm \frac{1}{2}$

$\therefore x = 0$ or $x = -1$

$\therefore$ The x-intercepts are $x = 0$ and $x = -1$

(OR $-3x^2 - 3x = 0$; $3x^2 + 3x = 0$; $3x(x + 1) = 0$ $\therefore x = 0$ or $x = -1$)

If $x = 0$, then $y = 0$

$\therefore$ The y-intercept : $y = 0$

(4) $y = \frac{1}{2}x^2 - 4x + 6$

$$y = \frac{1}{2}x^2 - 4x + 6 = \frac{1}{2}(x^2 - 8x) + 6 = \frac{1}{2}(x - 4)^2 - 8 + 6 = \frac{1}{2}(x - 4)^2 - 2$$

$\therefore$ vertex: $(4, -2)$ and axis of symmetry: $x = 4$

If $y = 0$, then $0 = \frac{1}{2}(x - 4)^2 - 2$; $\frac{1}{2}(x - 4)^2 = 2$; $x = 4 \pm 2$ $\therefore x = 6$ or $x = 2$

$\therefore$ The x-intercepts are $x = 6$ and $x = 2$

(OR $\frac{1}{2}x^2 - 4x + 6 = 0$; $x^2 - 8x + 12 = 0$; $(x - 6)(x - 2) = 0$ $\therefore x = 6$ or $x = 2$)

If $x = 0$, then $y = 6$

$\therefore$ The y-intercept : $y = 6$

#14 Find the value of $a + p + q$ for the following quadratic functions

(1) $y = -3x^2 + 4x - a + 1 = a(x + p)^2 + q$

$$y = -3x^2 + 4x - a + 1 = -3\left(x^2 - \frac{4}{3}x\right) - a + 1 = -3\left(x - \frac{2}{3}\right)^2 + \frac{4}{3} - a + 1$$

$\therefore a = -3$, $p = -\frac{2}{3}$, $q = \frac{4}{3} - a + 1$ ($a + q = \frac{7}{3}$)

$\therefore a + p + q = \frac{7}{3} - \frac{2}{3} = \frac{5}{3}$

(2) $y = \frac{1}{2}x^2 - ax + 1 = \frac{1}{2}(x+2)^2 + p + q$

$y = \frac{1}{2}x^2 - ax + 1 = \frac{1}{2}(x^2 - 2ax) + 1 = \frac{1}{2}(x-a)^2 - \frac{1}{2}a^2 + 1$

$\therefore a = -2$, $p + q = -\frac{1}{2}a^2 + 1$

$\therefore a + p + q = -2 - \frac{1}{2}a^2 + 1 = -\frac{1}{2}a^2 - 1 = -\frac{1}{2}(-2)^2 - 1 = -2 - 1 = -3$

(3) $y = ax^2 - 2x + 3 = -2(x+p)^2 - q$

$y = ax^2 - 2x + 3 = a\left(x^2 - \frac{2}{a}x\right) + 3 = a\left(x - \frac{1}{a}\right)^2 - \frac{1}{a} + 3$

$\therefore a = -2$, $p = -\frac{1}{a} = \frac{1}{2}$, $q = \frac{1}{a} - 3 = -\frac{7}{2}$

$\therefore a + p + q = -5$

#15 Find an equation of the quadratic function with the following conditions

(1) Vertex : $(1, 2)$ and passes through a point $(0, 3)$

$y = a(x - p)^2 + q$; $y = a(x - 1)^2 + 2$ using vertex

Substitute $(0, 3)$; $3 = a(-1)^2 + 2 = a + 2$; $a = 1$

$\therefore y = (x - 1)^2 + 2$

(2) Axis of symmetry : $x = -1$ and passes through two points $(-3, -2)$, $(0, 4)$

$y = a(x - p)^2 + q$; $y = a(x + 1)^2 + q$ using the axis of symmetry : $x = -1$

Substitute $(-3, -2)$, $(0, 4)$; $-2 = a(-3 + 1)^2 + q$ and $4 = a(0 + 1)^2 + q$

So, $4a + q = -2$ and $a + q = 4$ Thus, $4a + (4 - a) = -2$ $\therefore a = -2$ and $q = 6$

$\therefore y = -2(x + 1)^2 + 6$

(3) Vertex is on the x-axis, axis of symmetry : $x = -1$, and passes through a point $(-3, -4)$

$y = a(x - p)^2 + q$; $y = a(x - p)^2$ ($\because$ Vertex is on x-axis $\Rightarrow q = 0$)

Axis of symmetry : $x = -1$ $\Rightarrow p = -1$

$\therefore y = a(x + 1)^2$

Substitute $(-3, -4)$; $-4 = a(-3 + 1)^2$; $a = -1$

$\therefore y = -(x + 1)^2$

(4) Passes through three points $(0, -3), (2, -1)$ and $(4, -6)$

$y = ax^2 + bx + c$

$(0, -3) \Rightarrow c = -3$

$(2, -1) \Rightarrow -1 = 4a + 2b + c \Rightarrow 4a + 2b = 2 \; ; 2a + b = 1$

$(4, -6) \Rightarrow -6 = 16a + 4b + c \Rightarrow 16a + 4b = -3$

$\quad 8a + 4b = 4$

$-) \; \underline{16a + 4b = -3}$

$\quad -8a \quad = 7 \; ; \; a = -\frac{7}{8} \; \therefore b = -2a + 1 = \frac{11}{4}$

$\therefore \; y = ax^2 + bx + c = -\frac{7}{8}x^2 + \frac{11}{4}x - 3 = -\frac{7}{8}(x^2 - \frac{8}{7} \cdot \frac{11}{4}x) - 3$

$= -\frac{7}{8}(x^2 - \frac{2 \cdot 11}{7}x) - 3 = -\frac{7}{8}\left(x - \frac{11}{7}\right)^2 + \frac{121}{49} \cdot \frac{7}{8} - 3 = -\frac{7}{8}\left(x - \frac{11}{7}\right)^2 + \frac{121}{56} - 3$

$= -\frac{7}{8}\left(x - \frac{11}{7}\right)^2 - \frac{47}{56}$

(5) Passes through the origin, $(4, -3)$, and $(-2, 6)$

$y = ax^2 + bx + c$

$(0, 0) \Rightarrow c = 0 \; \therefore y = ax^2 + bx$

$(4, -3) \Rightarrow -3 = 16a + 4b$

$(-2, 6) \Rightarrow 6 = 4a - 2b$

$\quad 16a + 4b = -3$

$+) \; \underline{8a - 4b = 12}$

$\quad 24a \quad = 9 \; ; \; a = \frac{3}{8} \; \therefore 2b = 4a - 6 = \frac{3}{2} - 6 = -\frac{9}{2} \; ; b = -\frac{9}{4}$

$\therefore \; y = ax^2 + bx = \frac{3}{8}x^2 - \frac{9}{4}x = \frac{3}{8}\left(x^2 - \frac{8}{3} \cdot \frac{9}{4}x\right) = \frac{3}{8}(x^2 - 6x)$

$\quad = \frac{3}{8}((x - 3)^2 - 9) = \frac{3}{8}(x - 3)^2 - \frac{27}{8}$

(6) Passes through $(-3, 0), (6, 0)$ and $(0, -6)$

Since x-intercepts are -3 and 6, $y = a(x + 3)(x - 6)$

$(0, -6) \Rightarrow -6 = a(0 + 3)(0 - 6) = -18a \; ; \; a = \frac{1}{3}$

$\therefore y = \frac{1}{3}(x + 3)(x - 6) = \frac{1}{3}(x^2 - 3x - 18) = \frac{1}{3}(x^2 - 3x) - 6$

$= \frac{1}{3}\left(x - \frac{3}{2}\right)^2 - \frac{9}{4} \cdot \frac{1}{3} - 6 = \frac{1}{3}\left(x - \frac{3}{2}\right)^2 - \frac{27}{4}$

(7) Passes through $(-4, 1), (-2, 0)$ **and** $(0, 3)$

$y = ax^2 + bx + c$

$(0, 3) \Rightarrow 3 = c \quad \therefore y = ax^2 + bx + 3$

$(-2, 0) \Rightarrow 0 = 4a - 2b + 3$

$(-4, 1) \Rightarrow 1 = 16a - 4b + 3 \, ; \, 16a - 4b = -2 \, ; \, 8a - 2b = -1$

$ 4a - 2b = -3$

$ -) \, \underline{8a - 2b = -1}$

$ -4a = -2 \, ; \, a = \frac{1}{2} \, \therefore 2b = 4a + 3 = 5 \, ; \, b = \frac{5}{2}$

$\therefore \quad y = \frac{1}{2}x^2 + \frac{5}{2}x + 3 = \frac{1}{2}(x^2 + 5x) + 3 = \frac{1}{2}\left(\left(x + \frac{5}{2}\right)^2 - \frac{25}{4}\right) + 3$

$ = \frac{1}{2}\left(x + \frac{5}{2}\right)^2 - \frac{25}{8} + 3 = \frac{1}{2}\left(x + \frac{5}{2}\right)^2 - \frac{1}{8}$

#16 Find equations for the following parabolas

(1)

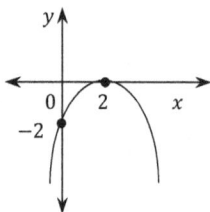

Since vertex is $(2, 0)$, $y = a(x - 2)^2 + 0$

$(0, -2) \Rightarrow -2 = 4a + 0 \, ; \, a = -\frac{1}{2}$

$\therefore y = -\frac{1}{2}(x - 2)^2$

(2)

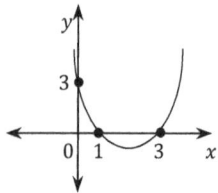

x-intercepts : $x = 1$ and $x = 3$

$\therefore y = a(x - 1)(x - 3)$

$(0, 3) \Rightarrow 3 = a(0 - 1)(0 - 3) = 3a \, ; \, a = 1$

$\therefore y = (x - 1)(x - 3) = x^2 - 4x + 3 = (x - 2)^2 - 4 + 3 = (x - 2)^2 - 1$

(3)

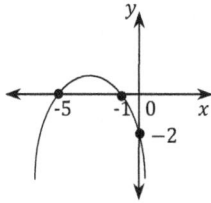

x-intercepts: $x = -1$ and $x = -5$

$\therefore y = a(x+1)(x+5)$

$(0,-2) \Rightarrow -2 = 5a$; $a = -\frac{2}{5}$

$\therefore y = -\frac{2}{5}(x+1)(x+5) = -\frac{2}{5}(x^2+6x+5) = -\frac{2}{5}(x^2+6x) - 2$

$= -\frac{2}{5}(x+3)^2 + \frac{18}{5} - 2 = -\frac{2}{5}(x+3)^2 + \frac{8}{5}$

(4)

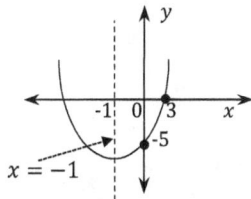

$x = -1$

$y = a(x+1)^2 + q$ using the axis of symmetry

$(3,0) \Rightarrow 0 = a(3+1)^2 + q$; $q = -16a$

$(0,-5) \Rightarrow -5 = a(0+1)^2 + q$; $q = -a - 5$

$\therefore -16a = -a - 5$; $a = \frac{1}{3}$

$\therefore q = -\frac{16}{3}$

Therefore, $y = \frac{1}{3}(x+1)^2 - \frac{16}{3}$

(5)

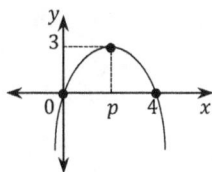

$y = a(x-p)^2 + 3$

$(0,0) \Rightarrow 0 = ap^2 + 3$

$(4,0) \Rightarrow 0 = a(4-p)^2 + 3 = a(16-8p+p^2) + 3 = 16a - 8ap + (ap^2+3)$

$\qquad = 16a - 8ap + 0 = 8a(2-p)$

Since $a \neq 0$, $p = 2$

Since $ap^2 + 3 = 0$, $4a + 3 = 0$; $a = -\frac{3}{4}$

$\therefore y = -\frac{3}{4}(x-2)^2 + 3$

(OR   since the x-intercepts are $x = 0$ and $x = 4$, the axis of symmetry is $x = 2$

(using the definition of the axis of symmetry).

So, the vertex is $(2, 3)$ Thus, $y = a(x - 2)^2 + 3$

$(0, 0) \Rightarrow 0 = 4a + 3$; $a = -\dfrac{3}{4}$

Therefore, $y = -\dfrac{3}{4}(x - 2)^2 + 3$

#17 Find the value of a for the following parabola

(1) The parabola $y = x^2 - ax + 2$ has $x = -2$ as its axis of symmetry.

$y = x^2 - ax + 2 = \left(x - \dfrac{a}{2}\right)^2 - \dfrac{a^2}{4} + 2$

$\therefore$ axis of symmetry : $x = \dfrac{a}{2} = -2$ $\therefore$ $a = -4$

(2) The parabola $y = -\dfrac{1}{2}x^2 + 4x - a + 1$ has its vertex on the x-axis.

$y = -\dfrac{1}{2}x^2 + 4x - a + 1 = -\dfrac{1}{2}(x^2 - 8x) - a + 1 = -\dfrac{1}{2}(x - 4)^2 + 8 - a + 1$

$= -\dfrac{1}{2}(x - 4)^2 - a + 9$

$\therefore$ vertex : $(4, -a + 9)$

Since the vertex is on the x-axis, $-a + 9 = 0$ $\therefore$ $a = 9$

(3) The distance between the two x-intercepts is 6 for a parabola $y = x^2 - 2x + a$.

$y = x^2 - 2x + a = (x - 1)^2 - 1 + a$

Since $x = 1$ is the axis of symmetry,

x-intercepts are 4 and -2 by the definition of the axis of symmetry.

$\therefore$ $(4, 0) \Rightarrow 0 = 9 - 1 + a$; $a = -8$

(OR $(-2, 0) \Rightarrow 0 = 9 - 1 + a$; $a = -8$)

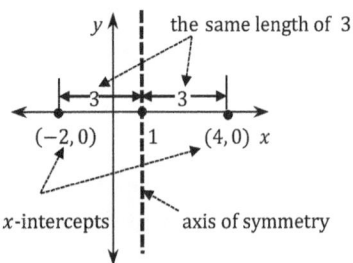

#18 Find the minimum value or maximum values for the following quadratic functions

(1) $y = x^2 - 4x + 5$

$y = (x-2)^2 - 4 + 5 = (x-2)^2 + 1$

Vertex : $(2, 1)$

The parabola opens upward.

$\therefore$ $y = 1$ is the minimum value at $x = 2$.

(2) $y = -2x^2 + 4x + 1$

$y = -2x^2 + 4x + 1 = -2(x^2 - 2x) + 1 = -2(x-1)^2 + 2 + 1 = -2(x-1)^2 + 3$

The parabola opens downward.

$\therefore$ $y = 3$ is the maximum value at $x = 1$.

(3) $y = -3(x+1)(x-3)$

Since $x = -1$ and $x = 3$ are the x-intercepts, we can get the axis of symmetry $x = 1$ using the same

distance between the axis of symmetry and x-intercepts.

$\therefore$ $y = 12$ is the maximum value at $x = 1$.

(OR $y = -3(x+1)(x-3) = -3(x^2 - 2x - 3) = -3((x-1)^2 - 1) + 9 = -3(x-1)^2 + 12$

Since parabola opens downward and $(1, 12)$ is the vertex,

$y = 12$ is the maximum value at $x = 1$.)

(4) $y = -x^2 + 4x - 4$

$y = -x^2 + 4x - 4 = -(x^2 - 4x) - 4 = -(x-2)^2 + 4 - 4 = -(x-2)^2$

Since the parabola opens downward and $(2, 0)$ is the vertex,

$y = 0$ is the maximum value at $x = 2$.

#19 Find the equation of the quadratic function with the following conditions

(1) The minimum value is 3 at $x = 1$ and passes through $(-2, 5)$.

$$y = a(x - p)^2 + q$$
$$y = a(x - 1)^2 + 3, \quad a > 0$$
$$(-2, 5) \implies 5 = 9a + 3 \; ; \; a = \frac{2}{9}$$
$$\therefore \; y = \frac{2}{9}(x - 1)^2 + 3$$

(2) The maximum value is 4 at $x = -1$ and passes through $(1, -8)$.

$$y = a(x + 1)^2 + 4, \quad a < 0$$
$$(1, -8) \implies -8 = 4a + 4 \; ; \; a = -3$$
$$\therefore \; y = -3(x + 1)^2 + 4$$

#20 Find the value of $a + b$ for the following quadratic function which has maximum or minimum value

(1) $y = -x^2 + 2ax + b$ has the maximum value 4 at $x = 1$.

$$y = -x^2 + 2ax + b = -(x^2 - 2ax) + b = -(x - a)^2 + a^2 + b$$
$$\text{Vertex} : (a, a^2 + b)$$
$$\therefore \; a = 1 \text{ and } a^2 + b = 4 \; ; \; b = 3$$
$$\therefore \; a + b = 4$$

$$(\text{OR} \; y = -(x - 1)^2 + 4 = -x^2 + 2x + 3 = -x^2 + 2ax + b \; \therefore 2 = 2a \text{ and } 3 = b$$
$$\therefore a + b = 1 + 3 = 4 \;)$$

(2) $y = 2x^2 - ax + b$ has the minimum value -3 at $x = -2$.

$$y = 2x^2 - ax + b = 2\left(x^2 - \frac{a}{2}x\right) + b = 2\left(x - \frac{a}{4}\right)^2 - \frac{a^2}{8} + b$$
$$\text{Vertex} : \left(\frac{a}{4}, -\frac{a^2}{8} + b\right)$$
$$\therefore \; \frac{a}{4} = -2 \; ; \; a = -8 \text{ and } -\frac{a^2}{8} + b = -3 \; ; b = 5$$
$$\therefore \; a + b = -3$$

$$(\text{OR} \; y = 2(x + 2)^2 - 3 = 2x^2 + 8x + 5 = 2x^2 - ax + b \; \therefore 8 = -a \text{ and } 5 = b$$
$$\therefore a + b = -8 + 5 = -3 \;)$$

(3) $y = ax^2 + 2x + b$ **has the maximum value** 3 **at** $x = 2$.

$$y = ax^2 + 2x + b = a\left(x^2 + \frac{2}{a}x\right) + b = a\left(x + \frac{1}{a}\right)^2 - \frac{1}{a} + b$$

Vertex : $\left(-\frac{1}{a}, -\frac{1}{a} + b\right)$, $a < 0$

$\therefore$ $-\frac{1}{a} = 2$; $a = -\frac{1}{2}$ and $-\frac{1}{a} + b = 3$; $b = 1$

$\therefore$ $a + b = \frac{1}{2}$

(OR $y = a(x - 2)^2 + 3$, $a < 0$

$\therefore$ $y = ax^2 - 4ax + 4a + 3 = ax^2 + 2x + b$ $\therefore -4a = 2$ and $4a + 3 = b$

$\therefore a + b = -\frac{1}{2} + 1 = \frac{1}{2}$)

(4) $y = ax^2 - bx + 2$ **has the maximum value** 3 **at** $x = -1$.

$$y = ax^2 - bx + 2 = a\left(x^2 - \frac{b}{a}x\right) + 2 = a(x - \frac{b}{2a})^2 - \frac{b^2}{4a} + 2$$

Vertex : $\left(\frac{b}{2a}, -\frac{b^2}{4a} + 2\right)$, $a < 0$

$\therefore$ $\frac{b}{2a} = -1$ and $-\frac{b^2}{4a} + 2 = 3$

$2a = -b$; $a = -\frac{b}{2}$ and $\frac{b^2}{4a} = -1$; $4a = -b^2$

So, $4\left(-\frac{b}{2}\right) = -b^2$; $b^2 - 2b = 0$; $b(b - 2) = 0$; $b = 0$ or $b = 2$

If $b = 0$, then $y = ax^2 - bx + 2 = ax^2 + 2$

Since $a < 0$, 2 is the maximum value for the parabola. But it's not true.

Therefore, $b \neq 0$. So, $b = 2$

$\therefore$ $a + b = -1 + 2 = 1$

(OR $y = a(x + 1)^2 + 3$, $a < 0$

$\therefore$ $y = ax^2 + 2ax + a + 3 = ax^2 - bx + 2$ $\therefore 2a = -b$ and $a + 3 = 2$

$\therefore a + b = -1 + 2 = 1$)

#21 One side of a rectangle is x inches. The perimeter and the area of a rectangle are 10 inches and y square inches, respectively. Find the maximum value of y.

Let x be the length of the rectangle. Then, $2x + 2 \cdot$ width$= 10$

So, width $= \frac{10-2x}{2} = 5 - x$

The area $y = x \cdot (5 - x) = -x^2 + 5x = -(x^2 - 5x) = -\left(x - \frac{5}{2}\right)^2 + \frac{25}{4}$

Therefore, the maximum value of the area is $\frac{25}{4}$ square inches when the length is $\frac{5}{2}$ inches.

#22 The sum of two numbers is 18. Find the maximum value of their product.

Since two numbers are x and $18 - x$,

the product of the numbers is $y = x(18 - x) = -x^2 + 18x = -(x^2 - 18x) = -(x - 9)^2 + 81$

Therefore, the maximum value is 81 when $x = 9$.

#23 The difference between two numbers is 10. Find the minimum value of their product.

Since the two numbers are x and $x - 10$,

the product of the numbers is $y = x(x - 10) = x^2 - 10x = (x - 5)^2 - 25$

When $x = 5$, the minimum value of the product is -25.

Therefore, the minimum value is -25 when the two numbers are 5 and -5.

#24 A ball is thrown upward from the top of a 5 foot table. After x seconds, the height of the ball from the ground is $y = -3x^2 + 12x + 5$. Find the maximum height from the ground the ball can reach.

$y = -3x^2 + 12x + 5 = -3(x^2 - 4x) + 5 = -3(x - 2)^2 + 12 + 5 = -3(x - 2)^2 + 17$

Vertex : $(2, 17)$

Therefore, after 2 seconds, the maximum height will be 17 feet.

Solutions Manual

CH. 1 ~ CH. 4

Statistics and Probability

Solutions for Chapter 1

#1 Refer to the bar graph below to answer the following questions.

Favorite Subjects

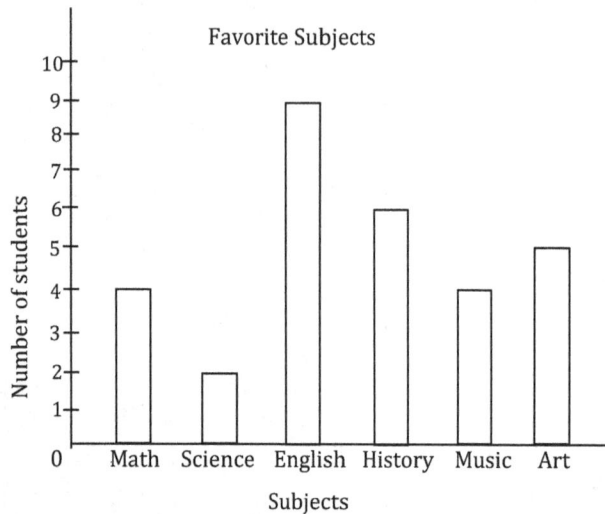

(1) Which subjects are the most and least favorite subjects?

The most : English

The least : Science

(2) Which two subjects are chosen by the same number of students?

Math and Music

(3) How many more students chose English than Math?

5 students

(4) What percentage of students chose history?

$4 + 2 + 9 + 6 + 4 + 5 = 30$

$\therefore \frac{6}{30} \times 100 = 20$ $\therefore 20\%$

#2   Present the following information in a circle graph.

Item	Price
Shoes	$45
Pants	$30
Food	$20
Snack	$15
Bag	$10

Since the total price is $120 (45 + 30 + 20 + 15 + 10 = 120),

$$\$45 \Rightarrow \frac{45}{120} \times 360° = 135°$$

$$\$30 \Rightarrow \frac{30}{120} \times 360° = 90°$$

$$\$20 \Rightarrow \frac{20}{120} \times 360° = 60°$$

$$\$15 \Rightarrow \frac{15}{120} \times 360° = 45°$$

$$\$10 \Rightarrow \frac{10}{120} \times 360° = 30°$$

Shopping Expenses

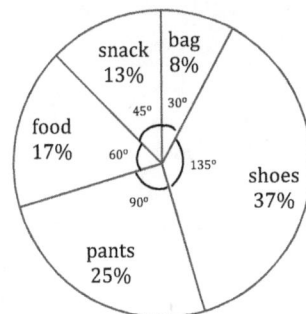

#3 According to the circle graph, how much did Nichole spend on each item for the current month?

Spending Expenses
Current Month
Total Income : $7,000

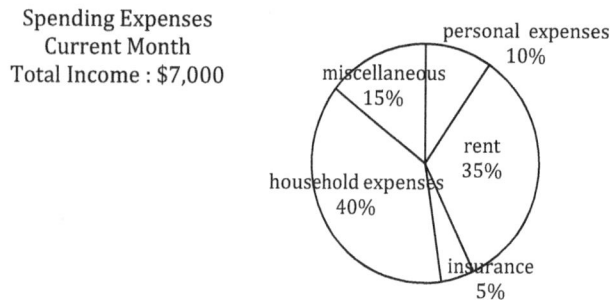

personal expenses
10%

miscellaneous
15%

rent
35%

household expenses
40%

insurance
5%

Personal expenses : $10\% \Rightarrow 7000 \times \frac{10}{100} = 700$ $\therefore$ $700

Rent : $35\% \Rightarrow 7000 \times \frac{35}{100} = 2450$ $\therefore$ $2450

Insurance : $5\% \Rightarrow 7000 \times \frac{5}{100} = 350$ $\therefore$ $350

Household expenses : $40\% \Rightarrow 7000 \times \frac{40}{100} = 2800$ $\therefore$ $2800

Miscellaneous : $15\% \Rightarrow 7000 \times \frac{15}{100} = 1050$ $\therefore$ $1050

#4 Complete the following table using the categorical data listed in the table

Category	Frequency	Relative Frequency	Central Angle(in degree)
A	50	0.208	$0.208 \times 360 = 74.88$
B	85	0.354	$0.354 \times 360 = 127.44$
C	65	0.271	$0.271 \times 360 = 97.56$
D	40	0.167	$0.167 \times 360 = 60.12$
Total	240	1.000	360

#5 Refer to the line graph below to answer the following questions.

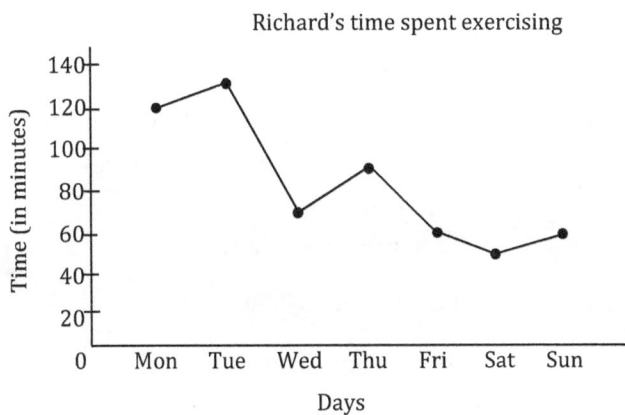

Richard's time spent exercising

(1) On which day did Richard exercise most?

Tuesday

(2) Which days did Richard exercise for the same amount of time?

Friday and Sunday

(3) Which two days show the greatest difference between Richard's time spent exercising?

Tuesday and Wednesday

Solutions for Chapter 2

#1 Make a stem-and-leaf plot for the following data.

43, 100, 50, 64, 73, 79, 81, 66, 55, 61, 101, 52, 55, 48, 64, 113, 77, 80, 81, 95, 53

```
 4 | 3  8
 5 | 0  2  3  5  5
 6 | 1  4  4  6
 7 | 3  7  9
 8 | 0  1  1
 9 | 5
10 | 0  1
11 | 3
```

#2 The following are the hourly wages in dollars of 20 workers. Arrange the data given below in a frequency table.

9.80 9.60 10.15 9.80 10.60 12.20 8.85 11.50 9.60 10.20

10.15 8.85 9.80 10.20 10.15 9.80 11.50 9.80 9.80 9.60

hourly wages	frequency
8.85	2
9.60	3
9.80	6
10.15	3
10.20	2
10.60	1
11.50	2
12.20	1

 # 3 The following distribution shows the number of days in a year each student in a class of 50 visited a doctor. Draw a relative frequency histogram.

Number of days	Number of students
1-5	4
6-10	6
11-15	11
16-20	13
21-25	9
26-30	7

Frequency	Relative frequency
4	$\frac{4}{50} = 0.08$
6	$\frac{6}{50} = 0.12$
11	$\frac{11}{50} = 0.22$
13	$\frac{13}{50} = 0.26$
9	$\frac{9}{50} = 0.18$
7	$\frac{7}{50} = 0.14$

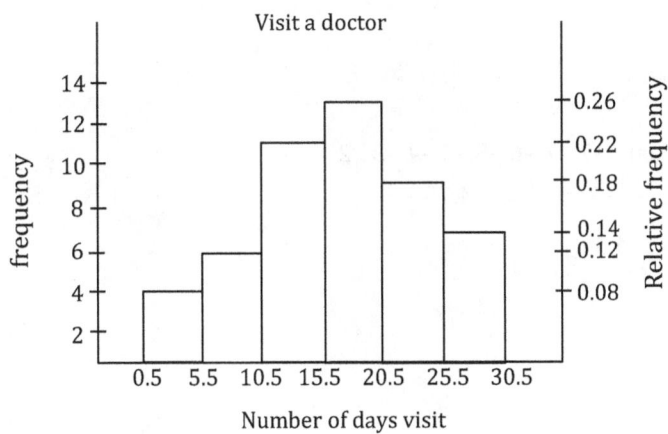

Visit a doctor

#4 Calculate the mean, median, mode, and range for the following sample measurements

(1) 3, 6, 7, 5, 4, 3, 2

$\Rightarrow$ List them in ascending order : 2, 3, 3, 4, 5, 6, 7

$\therefore$ Mean $= \frac{2+3+3+4+5+6+7}{7} = \frac{30}{7} = 4.286$ (; mean is the average of the seven measurements)

Median $= 4$ (4th value)

Mode $= 3$

Range $= 7 - 2 = 5$

(2) 9, 3, 5, 5, 2, 20, 4, 6

$\Rightarrow$ List them in ascending order : 2 3 4 5 5 6 9 20

$\therefore$ Mean $= \frac{2+3+4+5+5+6+9+20}{8} = \frac{54}{8} = 6.75$

Median $= 5$

Mode $= 5$

Range $= 20 - 2 = 18$

(3) 23.5, 31.2, 18.4, 35.4, 25

$\Rightarrow$ List them in ascending order : 18.4 23.5 25 31.2 35.4

$\therefore$ Mean $= \frac{18.4+23.5+25+31.2+35.4}{5} = \frac{133.5}{5} = 26.7$

Median $= 25$

No mode

Range $= 35.4 - 18.4 = 17$

(4) -4, -5, -7, -3, -4, -7, -2, -6

$\Rightarrow$ List them in ascending order : -7 -7 -6 -5 -4 -4 -3 -2

$\therefore$ Mean $= \frac{-7-7-6-5-4-4-3-2}{8} = \frac{-38}{8} = -4.75$

Median $= \frac{-4-5}{2} = \frac{-9}{2} = -4.5$

Mode $= -4$ and -7

Range $= -2 - (-7) = 5$

#5 Find the deviations, variance, and standard deviation for the following data

> 25, 30, 36, 38, 43, 56

$$\text{Mean} = \frac{25+30+36+38+43+56}{6} = \frac{228}{6} = 38$$

So, the deviations are $-13, -8, -2, \ 0, \ 5, \ 18$

The variance is $S^2 = \dfrac{(-13)^2+(-8)^2+(-2)^2+(0)^2+(5)^2+(18)^2}{6} = \dfrac{586}{6} = 97.67$

and the standard deviation is $s = \sqrt{S^2} = \sqrt{97.67} \approx 9.89$

#6 Refer to the data set below to answer the following questions.

> 27, 26, 27, 38, 23, 27, 39, 42, 38, 63, 34

(1) Determine the lower quartile, Q_L and upper quartile, Q_U .

⇒ List them in ascending order : 23, 26, 27, 27, 27, 34, 38, 38, 39, 42, 63
  Since median is $M = 34$,
  the lower quartile is $Q_L = 27$ and the upper quartile is $Q_U = 39$.

(2) Calculate the value of the interquartile range (IR).

The interquartile range is $IR = Q_U - Q_L = 39 - 27 = 12$.

(3) Draw a box plot.

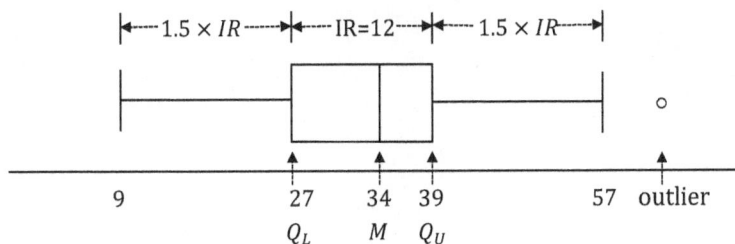

#7 Complete the following table and draw a line plot for the frequency distribution.

Measurement	Frequency	Relative Frequency (%)
36.4	3	$\frac{3}{45} = \frac{1}{15} = 6.67\% = 0.07$
42.5	6	$\frac{6}{45} = \frac{2}{15} = 13.33\% = 0.13$
48.1	4	$\frac{4}{45} = 8.89\% = 0.09$
53.2	6	$\frac{6}{45} = \frac{2}{15} = 13.33\% = 0.13$
55.8	8	$\frac{8}{45} = 17.78\% = 0.18$
64.7	9	$\frac{9}{45} = \frac{1}{5} = 20\% = 0.2$
66.3	6	$\frac{6}{45} = \frac{2}{15} = 13.33\% = 0.13$
72.9	3	$\frac{3}{45} = \frac{1}{15} = 6.67\% = 0.07$
total	45	$1 = 100\%$

Frequency distribution

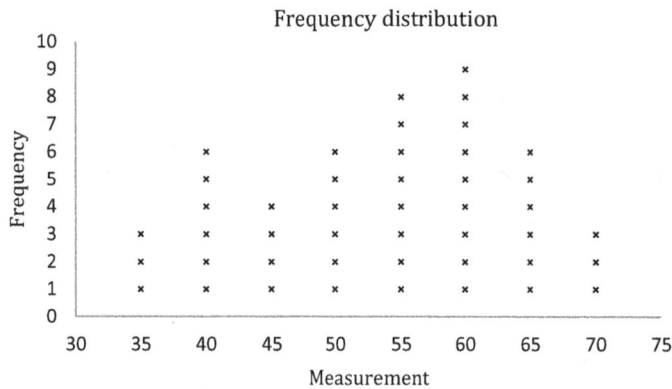

#8 Refer to the data set below to answer the following questions.

Age of teachers in a certain school.

43 40 55 47 37 36 52 40 28 26 42 40 36 45 39

(1) Draw a stem-and-leaf plot for the data.

```
2 | 6  8
3 | 6  6  7  9
4 | 0  0  0  2  3  5  7
5 | 2  5
```

(2) Complete the frequency table.

Age	Frequency	Relative Frequency (%)
26-30	2	$\frac{2}{15} = 13.3\%$
31-35	0	0%
36-40	7	$\frac{7}{15} = 46.7\%$
41-45	3	$\frac{3}{15} = 20\%$
46-50	1	$\frac{1}{15} = 6.7\%$
51-55	2	$\frac{2}{15} = 13.3\%$

(3) Draw a relative frequency histogram.

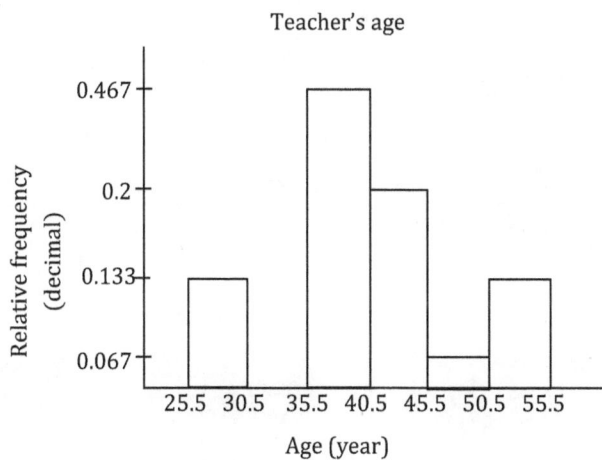

Teacher's age

(4) Find the mean, median, mode, and range.

Mean $= \frac{606}{15} = 40.4$; about 40 years old

Median $= 40$

Mode $= 40$

Range $= 55 - 26 = 29$

(5) Draw a line plot.

Teacher's age

(6) Draw a box plot.

Since median is 40, the box plot is

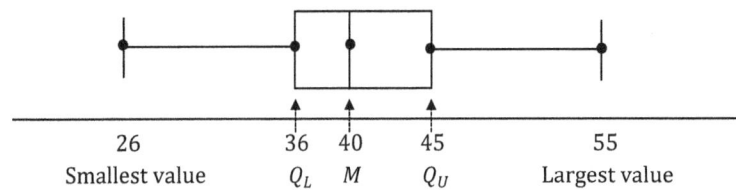

26	36	40	45	55
Smallest value	Q_L	M	Q_U	Largest value

Solutions for Chapter 3

#1 Identify all the sets. Mark o for a set or × for a non-set.

(1) A set of pretty girls in a school. ×

(2) A set of red apples. ×

(3) A set of natural numbers. o

(4) A set of small numbers. ×

(5) A set of famous singers. ×

(6) A set of even numbers. o

(7) A set of people who like math. ×

(8) A set of students whose heights are less than 5 feet in a class. o

(9) A set of 1-digit odd numbers. o

(10) The set of healthy foods in a store. ×

#2 Determine whether the following notations are true or false.

(1) $\{1, 2, 3\} = \{3, 1, 2\}$; true

(2) $\{1, 2, 3, 4, 5\} = \{x \mid x$ is a natural number less than 6.$\}$; true

(3) $\{x \mid x$ is a factor of 6$\} = \{1, 2, 3, 6\}$; true

(4) $\{x \mid x$ is a natural number less than 1.$\} = \{0\}$

false ($\because \{x \mid x$ is a natural number less than 1.$\} = \emptyset$)

(5) $\{0, 4, 8, 12, 16 \cdots\} = \{x \mid x$ is a multiple of 4.$\}$; true

(6) $\{1, 3, 5, 7, 9\} = \{x \mid x$ is an odd number less than 10.$\}$; true

(7) $\{x \mid 1 \leq x \leq 3, \ x$ is an integer.$\} = \{1, 2, 3\}$; true

(8) $\{1, 2, 3, 4\} = \{x \mid x$ is a prime number less than 5.$\}$

false ($\because \{x \mid x$ is a prime number less than 5.$\} = \{2, 3\})$

#3 State if the following sets are finite or infinite sets.

(1) $\{x \mid x$ is a factor of 20.$\}$; finite

(2) $\{x \mid x$ is a multiple of 2.$\}$; infinite

(3) $\{x \mid x$ is an even number.$\}$; infinite

(4) $\{x \mid 1 \leq x \leq 3, \ x$ is an odd number.$\}$; finite ($\because \emptyset$)

(5) $\{x \mid x^2 + 1 = 0, \ x$ is a real number.$\}$; finite ($\because \emptyset$)

(6) $\{x \mid x$ is an odd number bigger than 10.$\}$; infinite

#4 Find the value of $n(A) + n(B)$ for the following sets A and B

(1) $A = \{x \mid x \text{ is a factor of } 10.\}$, $B = \{x \mid x \text{ is an even number less than } 10.\}$

$A = \{1, 2, 5, 10\}$, $B = \{2, 4, 5, 8\}$ $\therefore n(A) + n(B) = 4 + 4 = 8$

(2) $A = \{x \mid 1 \leq x \leq 5,\ x \text{ is an odd number.}\}$, $B = \{0\}$

$A = \{1, 3, 5\}$, $B = \{0\}$ $\therefore n(A) + n(B) = 3 + 1 = 4$

(3) $A = \{x \mid x \text{ is a natural number less than } 1.\}$, $B = \{x \mid 3 < x < 4,\ x \text{ is a natural number.}\}$

$A = \emptyset$, $B = \emptyset$ $\therefore n(A) + n(B) = 0 + 0 = 0$

(4) $A = \{1, 2, 3, 4\}$, $B = \{2a + 1 \mid a \in A\}$

$A = \{1, 2, 3, 4\}$, $B = \{3, 5, 7, 9\}$ $\therefore n(A) + n(B) = 4 + 4 = 8$

#5 Find the value of $a + b$ for the following sets A and B

(1) $A = \{1, 2, a + 3\}$ and $B = \{2, 5, b + 1\}$, $A \subset B$ and $B \subset A$

$A \subset B$ and $B \subset A$ $\Rightarrow$ $A = B$

$\therefore$ $a + 3 = 5$ and $b + 1 = 1$

Therefore, $a = 2$ and $b = 0$. Hence, $a + b = 2$

(2) $A = \{3, 4, a + 1\}$ and $B = \{a + 2, b, 4\}$, $A \subset B$ and $B \subset A$

$A \subset B$ and $B \subset A$ $\Rightarrow$ $A = B$

$\therefore$ $3 \in B$ So, $3 = a + 2$ or $3 = b$

If $3 = a + 2 \Rightarrow a + 1 = b$

If $3 = b \Rightarrow a + 1 = a + 2$ $\Rightarrow$ impossible.

Thus, $3 = a + 2$ $\therefore a = 1$ and $b = a + 1 = 2$

Therefore, $a + b = 1 + 2 = 3$

#6 Find the number of subsets and proper subsets for the following sets A

(1) $A = \{x \mid x \text{ is a factor of } 15.\}$

$A = \{1, 3, 5, 15\}$ So, $n(A) = 4$

$\therefore$ The number of subsets is $2^4 = 16$ and the number of proper subsets is $2^4 - 1 = 15$.

(2) $A = \{x \mid x$ **is an even number less than 8.**$\}$

    $A = \{2, 4, 6\}$ So, $n(A) = 3$

    $\therefore$ The number of subsets is $2^3 = 8$ and the number of proper subsets is $2^3 - 1 = 7$.

#7 $A = \{x \mid x$ **is a factor of 12.**$\}$ **and** $B = \{x \mid x$ **is a factor of 6.**$\}$

How many number of subsets which include all the elements of B **are in the subsets of** A?

    $A = \{1, 2, 3, 4, 6, 12\}$ and $B = \{1, 2, 3, 6\}$

    $\therefore$ $2^{6-4} = 2^2 = 4$

#8 **How many number of subsets which include the element** a **but not the elements** b **and** c **are in the subsets of** $A = \{a, b, c, d, e, f\}$ **?**

    $2^{6-1-2} = 2^3 = 8$.

    It is same as the number of subsets of $\{d, e, f\}$ which is not including the elements a, b and c.

#9 **Find the number of a set** A **which satisfies the conditions for (1), (3), and (4). For (2), find a set** A.

(1) $\{1\} \subset A \subset \{1, 2, 3\}$

    A is the subset of $\{1, 2, 3\}$ including the element 1. $\therefore$ $2^{3-1} = 4$

(2) $\{2, 3\} \subset A \subset \{2, 3, 4, 5, 6\}$ **and** $n(A) = 3$

    $2 \in A$ and $3 \in A$

    Since $n(A) = 3$, $A = \{2, 3, 4\}$ or $\{2, 3, 5\}$ or $\{2, 3, 6\}$

(3) $A \subset \{x \mid x$ **is a natural number less than 5.**$\}$ **and** A **has at least one even number.**

    $A \subset \{1, 2, 3, 4\}$

    From the number of subsets of $\{1, 2, 3, 4\}$, exclude the number of subsets of $\{1, 3\}$.

    $\therefore$ $2^4 - 2^2 = 16 - 4 = 12$

(4) $A \subset \{x \mid x$ **is a factor of 20.**$\}$ **and** $(1 \in A$ **or** $2 \in A)$

    $A \subset \{1, 2, 4, 5, 10, 20\}$

    From the number of subsets of $\{1, 2, 4, 5, 10, 20\}$, exclude the number of subsets of $\{4, 5, 10, 20\}$.

    $\therefore$ $2^6 - 2^4 = 64 - 16 = 48$

#10 Find the value of $p + q$.

(1) **The number of subsets of A is 64 and $n(A) = p$.**

The number of proper subsets of B is 7 and $n(B) = q$.

$64 = 2^6$ $\therefore n(A) = 6$ $\therefore p = 6$

$7 = 2^3 - 1$ $\therefore n(B) = 3$ $\therefore q = 3$

Therefore, $p + q = 6 + 3 = 9$.

(2) $A \subset \{x \mid 1 \leq x \leq p + q, \ x \text{ is a natural number.}\}$,

$(p \in A \text{ and } q \in A)$, and $n(A) = 32$, where $p + q > 2$

$2^{p+q-2} = 32 = 2^5$ $\therefore p + q - 2 = 5$ Therefore, $p + q = 7$

(3) $A \subset \{x \mid 1 \leq x \leq p + q, \ x \text{ is a natural number.}\}$,

$(p \in A \text{ and } p + q \in A)$, $1 \notin A$, and $n(A) = 32$, where $p + q > 3$

$2^{p+q-3} = 2^5$ $\therefore p + q - 3 = 5$ Therefore, $p + q = 8$

#11 Find the intersection of the following sets

(1) $A = \{x \mid x \text{ is a factor of 6.}\}$, $B = \{x \mid 1 \leq x \leq 10, \ x \text{ is an even number.}\}$

$A = \{1, 2, 3, 6\}$, $B = \{2, 4, 6, 8, 10\}$ $\therefore A \cap B = \{2, 6\}$

(2) $A = \{1, 2, 3, 4, 5\}$, $B = \{x \mid x = a + 1, \ a \in A\}$

$B = \{2, 3, 4, 5, 6\}$ $\therefore A \cap B = \{2, 3, 4, 5\}$

(3) $A = \{x \mid x \text{ is a multiple of 3.}\}$, $B = \{x \mid x \text{ is a factor of 12.}\}$

$A = \{0, 3, 6, 9, \cdots \cdots\}$, $B = \{1, 2, 3, 4, 6, 12\}$ $\therefore A \cap B = \{3, 6, 12\}$

(4) $A = \{x \mid x \text{ is an even number.}\}$, $B = \{x \mid x \text{ is an odd number.}\}$

$A \cap B = \emptyset$

#12 Find the set A which satisfies the following conditions

(1) $B = \{1, 2, 3\}$, $A \cup B = \{1, 2, 3, 4, 5\}$, $A \cap B = \emptyset$

$A = \{4, 5\}$

(2) $B = \{1, 2, 3, 4\}$, $A \cup B = \{1, 2, 3, 4, 5, 6\}$, $A \cap B = \{1, 2\}$

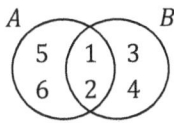

$\therefore A = \{1, 2, 5, 6\}$

(3) $B = \{1, 2, 3\}$, $A \cup B = \{1, 2, 3, 4, 5\}$, $n(A \cap B) = 2$

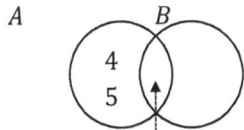

$\{1, 2\}$ or $\{1, 3\}$ or $\{2, 3\}$

$\therefore A = \{1, 2, 4, 5\}$ or $\{1, 3, 4, 5\}$ or $\{2, 3, 4, 5\}$

(4) $A = \{a, a+1, a+2\}$, $B = \{3, 4, 5, 6, 7\}$, $A \cap B = \{3, 4\}$

Case 1. $a = 3$, $a + 1 = 4$ $\Rightarrow$ $A = \{3, 4, 5\}$ $\Rightarrow$ $A \cap B = \{3, 4, 5\}$; false

Case 2. $a + 1 = 3$, $a + 2 = 4$ $\Rightarrow$ $A = \{2, 3, 4\}$ $\Rightarrow$ $A \cap B = \{3, 4\}$; true

$\therefore A = \{2, 3, 4\}$

#13 Find the number of a set A with the following conditions

(1) $B = \{x \mid x \text{ is a factor of } 6.\}$, $C = \{x \mid x \text{ is a factor of } 18.\}$, $A \cap B = B$, $A \cup C = C$

$B = \{1, 2, 3, 6\}$, $C = \{1, 2, 3, 6, 9, 18\}$

$A \cap B = B \Rightarrow B \subset A$

$A \cup C = C \Rightarrow A \subset C$

$\therefore B \subset A \subset C$

$\therefore$ The number of A is the same as the number of subsets of C including all the elements 1, 2, 3, 6 of B.

Since $n(C) = 6$ and $n(B) = 4$, $n(A) = 2^{6-4} = 4$

(2) For a set B, $n(B) = 5$, $n(A \cap B) = 3$, and $n(A \cup B) = 10$

Since $n(A \cup B) = n(A) + n(B) - n(A \cap B)$, $10 = n(A) + 5 - 3$.

$\therefore n(A) = 8$

(3) For a set B, $A \cap B = \emptyset$, $n(B) = 7$, and $n(A \cup B) = 15$

Since $n(A \cup B) = n(A) + n(B) - n(A \cap B)$, $15 = n(A) + 7 - 0$.

$\therefore n(A) = 8$

(4) For a set B, $n(A \cup B) = 20$, $n(A \cap B) = 5$, and $n(B - A) = 8$

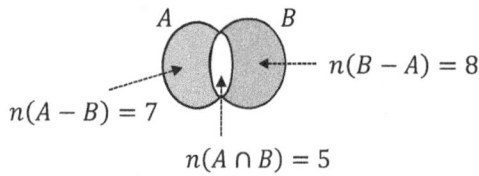

$$\therefore n(A) = 7 + 5 = 12$$

(5) For a fixed set $U = \{a, b, c, d, e, f, g\}$ and a set B,

$A - B = \{a, b\}$, $B - A = \{c, d\}$, and $(A \cup B)^C = \{f\}$

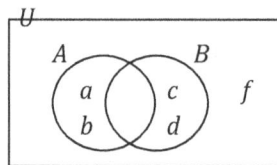

$\therefore A \cap B = \{e, g\}$ $\quad \therefore A = \{a, b, e, g\}$

$\therefore n(A) = 4$

(6) For two sets B and C, $n(B \cup C) = 10$, $n(B \cap C) = 3$, $n(C) = 5$, and $A \subset B$, $A \cap C = \emptyset$

Since $A \subset B$ and $A \cap C = \emptyset$, $C \subset A^C$ or $C \subset B^C$

If C is in B^C, then $B \cap C = \emptyset$.

But, $n(B \cap C) = 3$. So, C must be in A^C.

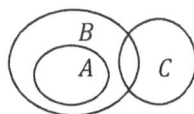

Since $n(B \cup C) = n(B) + n(C) - n(B \cap C)$, $10 = n(B) + 5 - 3$ $\therefore n(B) = 8$.

Since $A \subset B$, the number of A is the same as the number of subsets of B which is not including all the elements of $B \cap C$.

$\therefore 2^{8-3} = 2^5 = 32$.

#14 Find the sets $A - B$ and $B - A$ for the following sets A and B.

(1) $A = \{1, 2, 3, 4, 5\}$, $B = \{x \mid x$ is a factor of 4.$\}$

$B = \{1, 2, 4\}$ ∴ $A - B = \{3, 5\}$ and $B - A = \emptyset$

(2) $A = \{1, 2, 3, 4, 5, 6\}$, $B = \{x \mid 1 \leq x \leq 7, \ x$ is an odd number.$\}$

$B = \{1, 3, 5, 7\}$ ∴ $A - B = \{2, 4, 6\}$ and $B - A = \{7\}$

(3) For a fixed set $U = \{x \mid x$ is a factor of 20.$\}$,

$A \subset U, \ B \subset U, \ A \cap U = \{2, 5, 10\}, \ A \cap B = \{5, 10\}, \ (A \cup B)^C = \{1\}$

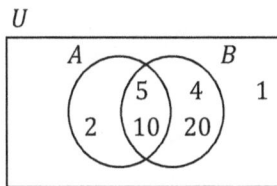

$U = \{1, 2, 4, 5, 10, 20\}$

∴ $A - B = \{2\}$ and $B - A = \{4, 20\}$

#15 Solve the following operations for any subsets A and B of the fixed set U.

(1) $A^C = U - A$

(2) $(A^C)^C = A$

(3) $U - A^C = A$

(4) $A - A^C = A \cap (A^C)^C = A \cap A = A$

(5) $(A \cup A^C) - (A \cap A^C) = U - \emptyset = U$

(6) $A \cap B^C = A - B$

(7) $A^C \cap B^C = (A \cup B)^C = U - (A \cup B)$

(8) $A - B$ when $A \cap B = A$

∴ $A - B = \emptyset$

(9) $A - B$ when $A \cap B = \emptyset$

∴ $A - B = A$

(10) $A \cap B$ when $A - B = A$

$A - B = A \cap B^C = A \quad \therefore A \subset B^C$

$\therefore A \cap B = \emptyset$

(11) $A^C - B^C$ when $A \subset B$

$A^C - B^C = A^C \cap (B^C)^C = A^C \cap B = B - A$

Solutions for Chapter 4

#1 Find the sample space for the following

  (1) A spin of a spinner marked 0, 1, 2, 3

  $\{0, 1, 2, 3\}$

  (2) A toss of one coin and a spin of a spinner marked 0, 1, 2, 3

  $\{$H0, H1, H2, H3, T0, T1, T2, T3$\}$

  (3) Two coins tossed once

  $\{$HH, HT, TH, TT$\}$

  (4) A pair of dice tossed once

  $\{11, 12, \cdots, 16, 21, 22, \cdots 26, \cdots\cdots, 61, 62, \cdots, 66\}$

  (5) A toss of one coin and a toss of an ordinary die.

  $\{$H1, H2, $\cdots$, H6, T1, T2, $\cdots$, T6 $\}$

#2 Think about tossing one coin and one ordinary die. Find the probability of the event

  (1) E : two heads occur.

  $P(E) = P(\emptyset) = 0$

  (2) F : a head and an even number occurs.

  $F = \{(H, 2), (H, 4), (H, 6)\}$

  Since each of the points is weighted $\frac{1}{12}$, $P(F) = \frac{3}{12} = \frac{1}{4}$.

  (3) G : an odd or even number occurs.

  $G = \{(H, 1), (H, 3), (H, 5), (T, 1), (T, 3), (T, 5), (H, 2), (H, 4), (H, 6), (T, 2), (T, 4), (T, 6)\}$

  $\therefore P(G) = \frac{12}{12} = 1$.

#3 Find the probabilities for the indicated sample spaces derived from the random experiment of drawing one card from a full deck.

  (1) $S = \{$red , black$\}$

  $P(\text{red}) = P(\text{black}) = \frac{26}{52}$

  (2) $S = \{$ace or picture card , otherwise$\}$

  $P(\text{ace or picture card}) = \frac{16}{52}$, $P(\text{otherwise}) = \frac{36}{52}$

#4 One ball is drawn from a bag containing three red balls marked 1, 2, 3 ; four blue balls marked 1, 2, 3, 4 ; and two yellow balls marked 1, 2. Find the probability for the indicated sample spaces from this random experiment.

    (1) $S = \{red,\ blue,\ yellow\}$

        $P(red) = \frac{3}{9} = \frac{1}{3},\ \ P(blue) = \frac{4}{9},\ \ P(yellow) = \frac{2}{9}$

    (2) $S = \{even,\ odd\}$

        $P(even) = \frac{4}{9},\ \ P(odd) = \frac{5}{9}$

#5 Three coins are tossed at once. Find $P(E \cap F)$.

    (1) Let E be the event "coins match" and let F be the event "not more than one head".

        $E = \{(H, H, H), (T, T, T)\},\ \ \ F = \{(H, T, T), (T, H, T), (T, T, H), (T, T, T)\}$ (1 head or 0 head)

        Since $E \cap F = \{ (T, T, T)\},\ \ P(E \cap F) = \frac{1}{8}$

    (2) Let E be the event "coins match" and let F be the event "not more than three heads".

        $E = \{(H, H, H), (T, T, T)\},$

        $F = \{(H, H, H), (H, T, T), (T, H, T), (T, T, H), (H, H, T), (T, H, H), (H, T, H), (T, T, T)\}$

            (3 heads, 2, heads, 1 head, or 0 head)

        Since $E \cap F = \{ (H, H, H), (T, T, T)\},\ \ P(E \cap F) = \frac{2}{8} = \frac{1}{4}$

    (3) Let E be the event "coins match" and let F be the event "at least two heads".

        $E = \{(H, H, H), (T, T, T)\},$

        $F = \{(H, H, T), (T, H, H), (H, T, H), (H, H, H)\}$ (2 heads or 3 heads)

        Since $E \cap F = \{ (H, H, H)\},\ \ P(E \cap F) = \frac{1}{8}$

    (4) Let E be the event "coins match" and let F be the event "at least one head".

        $E = \{(H, H, H), (T, T, T)\},$

        $F = \{(H, T, T), (T, H, T), (T, T, H), (H, H, T), (T, H, H), (H, T, H), (H, H, H)\}$ (1 head, 2 heads or 3 heads)

        Since $E \cap F = \{ (H, H, H)\},\ \ P(E \cap F) = \frac{1}{8}$

(5) Let E be the event "head on first toss" and let F be the event "tail on second toss".

$E = \{(H, H, H), (H, H, T), (H, T, H), (H, T, T)\},$

$F = \{(H, T, H), (H, T, T), (T, T, H), (T, T, T)\}$

Since $E \cap F = \{ (H, T, H), (H, T, T)\},\quad P(E \cap F) = \frac{2}{8} = \frac{1}{4}$

#6 For the following experiments, one toss is made. Find the probabilities indicated.

(1) Two coins, E : "at most one head", F : "no tails". Find $P(E \cup F)$.

$E = \{(H, T), (T, H), (T, T)\}$

$F = \{(H, H)\}$

Since $E \cap F = \emptyset,\quad P(E \cup F) = P(E) + P(F) - P(E \cap F) = \frac{3}{4} + \frac{1}{4} - 0 = \frac{4}{4} = 1$

(2) Three coins, E : "at least two heads", F : "only one tail". Find $P(E \cup F)$.

$E = \{(H, H, H), (H, H, T), (H, T, H)(T, H, H)\}$

$F = \{(T, H, H), (H, T, H), (H, H, T)\}$

Since $E \cap F = \{(H, H, T), (H, T, H)(T, H, H)\},$

$P(E \cup F) = P(E) + P(F) - P(E \cap F) = \frac{4}{8} + \frac{3}{8} - \frac{3}{8} = \frac{4}{8} = \frac{1}{2}$

(3) Two dice, $E = \{(1, 2)\}$, $F = \{(3, 4)\}$. Find $P(E \cup F)$.

$P(E \cup F) = P(E) + P(F) - P(E \cap F) = \frac{1}{36} + \frac{1}{36} - 0 = \frac{2}{36} = \frac{1}{18}$

(4) Two dice, E : "The sum of marked numbers is 6". Find $P(E)$.

Since $E = \{(1, 5), (2, 4), (3, 3), (4, 2), (5, 1)\},\ P(E) = \frac{5}{36}$

(5) Two dice, E : "sum $\leq$ 10". Find $P(E)$.

Since E^C is the event that " sum $\geq$ 11", $E^C = \{(5, 6), (6, 5), (6, 6)\}$

Since $(E^C) = \frac{3}{36} = \frac{1}{12}$, $P(E) = 1 - P(E^C) = 1 - \frac{1}{12} = \frac{11}{12}$

#7 **Find the number of different arrangements (permutations) for the following events**

(1) Scheduling seven different classes in seven periods.

$7! = 7 \cdot 6 \cdot 5 \cdot 4 \cdot 3 \cdot 2 \cdot 1 = 5{,}040$

There are 5,040 different arrangements for the classes.

(2) Creating the batting order for a baseball team consisting of 9 players.

$9! = 9 \cdot 8 \cdot 7 \cdot 6 \cdot 5 \cdot 4 \cdot 3 \cdot 2 \cdot 1 = 362{,}880$

There are 362,880 possible batting orders.

#8 **A coin is flipped twice. What is the conditional probability that both coins are heads, given that the first coin is a head?**

$$E = \{(H, H)\}, \quad F = \{(H, H), (H, T)\}, \quad E \cap F = \{(H, H)\}$$

$$\therefore \ P(E \backslash F) = \frac{P(E \cap F)}{P(F)} = \frac{\frac{1}{4}}{\frac{2}{4}} = \frac{1}{2}$$

#9 **A bag contains 5 red, 7 white, and 10 black balls. A ball is chosen at random from the bag, and it is noted that it is not one of the white balls. What is the conditional probability that it is red?**

Let R be the event that the red ball is chosen and let W^C be the event that it is not white.

Then, $P(R \backslash W^C) = \frac{P(R \cap W^C)}{P(W^C)}$

Since $R \cap W^C = R$, $P(R \cap W^C) = P(R) = \frac{5}{22}$

Therefore, $P(R \backslash W^C) = \frac{\frac{5}{22}}{\frac{15}{22}} = \frac{5}{15} = \frac{1}{3}$

#10 A box contains 10 apples and 15 pears. The fruits to be chosen are selected at random. Find the probability that

(1) The first two fruits chosen are apples.

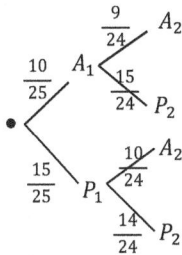

We have the event apple on the first and apple on the second.

So, the event is $A_1 \cap A_2$.

Since $P(A_1 \backslash A_2) = \frac{P(A_1 \cap A_2)}{P(A_2)}$, $P(A_1 \cap A_2) = P(A_2)P(A_1 \backslash A_2)$.

Since $P(A_2) = \frac{10}{25}$ and one apple is removed, $P(A_1 \backslash A_2) = \frac{9}{24}$.

Therefore, $P(A_1 \cap A_2) = \frac{10}{25} \cdot \frac{9}{24} = \frac{3}{20}$

(2) The second fruit chosen is an apple.

Event : $\{(A_1, A_2), (P_1, A_2)\}$

$P(A_2) = P(A_1 \cap A_2) + P(P_1 \cap A_2) = \frac{3}{20} + P(P_1 \cap A_2)$

Since $P(A_2 \backslash P_1) = \frac{P(A_2 \cap P_1)}{P(P_1)}$, $P(A_2 \cap P_1) = P(P_1)P(A_2 \backslash P_1)$.

So, $P(P_1 \cap A_2) = P(A_2 \cap P_1) = P(P_1)P(A_2 \backslash P_1) = \frac{15}{25} \cdot \frac{10}{24} = \frac{1}{4}$

Therefore, $P(A_2) = \frac{3}{20} + \frac{1}{4} = \frac{8}{20} = \frac{2}{5}$

(3) Given that the second fruit chosen is an apple, the first fruit chosen is also an apple.

$P(A_1 \backslash A_2) = \frac{P(A_1 \cap A_2)}{P(A_2)} = \frac{\frac{3}{20}}{\frac{2}{5}} = \frac{3}{8}$

#11 Think about tossing a coin and an ordinary dice. Let E be the event " H on coin " and let F be the event " 2 on dice ". Find $P(E \cup F), P(E \cap F),$ and determine if E and F are dependent or independent.

$S = \{H1, H2, \cdots, H6, T1, T2, \cdots, T6\}$

$E = \{H1, H2, H3, H4, H5, H6\}, \quad F = \{H2, T2\}$

$E \cup F = \{H1, H2, H3, H4, H5, H6, T2\}, \quad E \cap F = \{H2\}$

$\therefore \ P(E \cup F) = \frac{7}{12}, \ P(E \cap F) = \frac{1}{12}$

Since $P(E) = \frac{6}{12} = \frac{1}{2}$ and $P(F) = \frac{2}{12} = \frac{1}{6}, \ P(E) \cdot P(F) = \frac{1}{2} \cdot \frac{1}{6} = \frac{1}{12}.$

$\therefore \ P(E \cap F) = P(E) \cdot P(F)$

Therefore, E and F are independent.

#12 Think about tossing a nickel and a dime. Let E be the event " coins match " , let F be the event " nickel falls on heads " , and let G be the event " at least one head shows ". Determine which events are independent.

$E = \{(H, H), (T, T)\}, \quad F = \{(H, H), (H, T)\}, \quad G = \{(H, H), (H, T), (T, H)\}$

$E \cap F = \{(H, H)\}, \quad E \cap G = \{(H, H)\}, \quad F \cap G = \{(H, H), (H, T)\}$

$\therefore \ P(E) = \frac{2}{4} = \frac{1}{2}, \ P(F) = \frac{2}{4} = \frac{1}{2}, \ P(G) = \frac{3}{4}, \ P(E \cap F) = \frac{1}{4}, \ P(E \cap G) = \frac{1}{4}, \ P(F \cap G) = \frac{2}{4} = \frac{1}{2}$

Since $P(E) \cdot P(F) = \frac{1}{4}$ and $P(E \cap F) = \frac{1}{4}, \ P(E \cap F) = P(E) \cdot P(F)$

Therefore, E and F are independent.

Since $P(E \cap G) = \frac{1}{4}$ and $P(E) \cdot P(G) = \frac{1}{2} \cdot \frac{3}{4} = \frac{3}{8}, \ P(E \cap G) \neq P(E) \cdot P(G)$

Therefore, E and G are dependent.

Since $P(F \cap G) = \frac{1}{2}$ and $P(F) \cdot P(G) = \frac{1}{2} \cdot \frac{3}{4} = \frac{3}{8}, \ P(F \cap G) \neq P(F) \cdot P(G)$

Therefore, F and G are dependent.

Index (for Functions)

Index (for Statistics and Probability)

www.ingramcontent.com/pod-product-compliance
Lightning Source LLC
Chambersburg PA
CBHW061754210326
41518CB00036B/2335